iesel Engine

ZER Diesel
ZER Diesel
ZER Diesel
ZER Diesel
ZER Diesel
ZER Diesel

SULZER Diesel
SULZER Diesel
SULZER Diesel
SULZER Diesel
SULZER Diesel
SULZER Diesel
SULZER Diesel
SULZER Diesel
SULZER Diesel
SULZER Diesel
SULZER Diesel
SULZER Diesel
SULZER Diesel
SULZER Diesel
SULZER Diesel
SULZER Diesel
SULZER Diesel
SULZER Diesel

From The Mountains To The Seas

The Sulzer Diesel Engine

Text by	Jack A Somer David T Brown (history)
Photographs by	Helmut Behling Wärtsilä NSD Switzerland Ltd
Concept by	Jürgen Gerdes
Published by	Wärtsilä NSD Switzerland Ltd DENON Publizistik AG

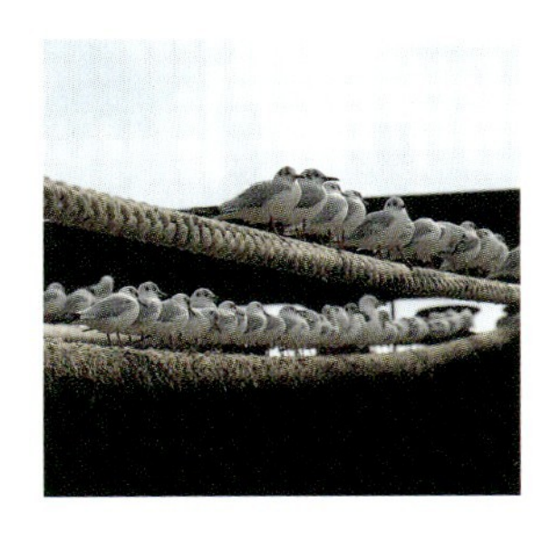

From The Mountains To The Seas

The Sulzer Diesel Engine

ISBN 3-9521561-0-8

Project responsibility	Jürgen Gerdes, Henggart, Switzerland
Project management	DENON Publizistik AG, Rapperswil, Switzerland
Author	Jack A Somer, Greenwich, Connecticut, USA David T Brown, Winterthur, Switzerland (history)
Photographs	Helmut Behling, Motive vom Norden, Bremen, Germany Wärtsilä NSD Switzerland Ltd
Cover design	Bodmer & Partner, Rapperswil, Switzerland
Design, layout, print and setting	Zürichsee Druckereien AG, Stäfa, Switzerland Design: Edith Camen Layout: Franziska Rose and Sherif Ademi

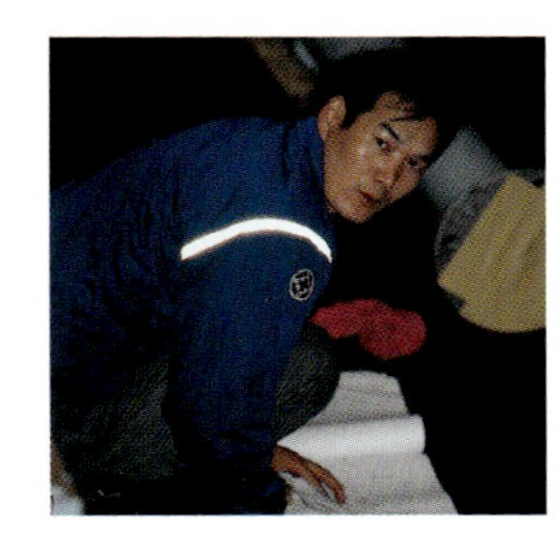

Contents

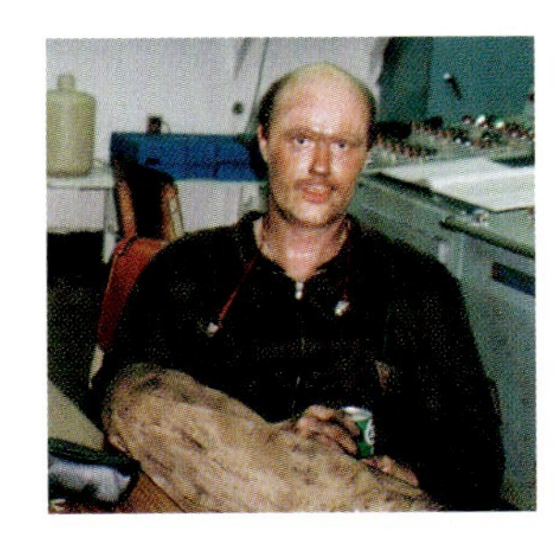

Foreword

More than a century ago, few people could have imagined that a small town in a land-locked European country would become home to one of the greatest of all technological inventions, the diesel engine. It was on 10 June 1898 that the first Sulzer diesel was started up in Winterthur, Switzerland, as a result of a very special relationship between Rudolf Diesel and the Sulzer family. On that date, the Sulzer diesel embarked on a successful voyage that has not lost any of its momentum.

One hundred years ago our forefathers had the bold vision that diesel technology would soon replace steam and become prime mover for a great variety of applications, the most significant being marine propulsion. With the thoroughness and dedication characteristic of the Swiss, Winterthur's engineers pushed diesel technology to one breakthrough and milestone after another. From the design of the first reversible engine, to the installation in 1997 of the most powerful diesel ever built, Sulzer diesel engines have made history.

Descending from the mountains to the seas, Sulzer marine diesels propel thousands of ocean-going vessels, and Sulzer became a household name with shipowners and shipyards, and even in university textbooks. The success of the Sulzer diesel, however, would have not been possible without another revolutionary concept. Recognising that large marine engines had to be built close to shipyards, Sulzer pioneered the idea of licensing their technology around the world, carrying this partnership philosophy to East Asia as early as 1917. Today we are grateful to our licensees who, over many years, have helped us make the Sulzer diesel engine what it is today.

Recently, time seems to run faster and faster. This applies to technology's developmental cycles as well as to the economic and corporate environment we live in. Change is the only constant, and "globalisation" is the key word for progress and survival. It is not surprising, therefore, that the Sulzer diesel engine has also been touched by this evolution, having recently become a prominent part of the world's leading large diesel engine group. The stage is set for another successful new century of the Sulzer brand.

Our heartfelt thanks go out to our forefathers, employees, partners and friends of the past, present, and future, for their great confidence in and everlasting support for the Sulzer diesel engine.

Peter G. Sulzer
Chairman of Wärtsilä NSD Switzerland

Yesterday

Between 1903 and the dawn of 1998, a grand total of 36,980 diesel engines, totalling 158.3 million horsepower, were built with the "Sulzer" nameplate on them. This is the story of those thousands of Sulzer diesel engines. As such, it is an absorbing story of gleaming steel pistons, grey cast cylinders, gargantuan sculptured crankshafts, mean effective pressures, impossible lubrication, celestial-level temperatures, and the limitless power of combustion. It is also the story of great ships, whose bow waves are ever crested with foam owing to the indomitable push of Sulzer diesel engines and, to a lesser extent, great choruses of diesel power stations sending electricity into homes and factories all over the world.

But, if the story of the Sulzer-brand diesel were to be told in technical terms alone, it would be only half a story. For the history of the Sulzer diesel engine, if it is anything, is the saga of the human quest for mastery over Rudolf Diesel's wildest dream, by many other men and women of science, engineering, and faith, who — with their feet planted solidly on the ground of internal combustion technology — have dared to reach for the combustive power of the stars.

Rudolf Diesel's invention is more than 100 years old. By any normal measure today, a 100-year-old invention, dependent upon 100-year-old technology, is a tired invention, an invention begging to be replaced. (If a pilot of a late-20th-century transoceanic airliner were ever to look out his window and see a puny petrol engine of the sort the Wright brothers used nearly a century ago, he would surely make ready to bail out.) But the diesel engine and its technology show no signs of being replaced. And that is largely because so many people, in so many disciplines — mathematics, mechanics, thermodynamics, metallurgy, combustion chemistry, electronics, computer modelling, fuel and lubricant technology, and above all diplomacy — have dedicated their lives to trying to make the Sulzer diesel engine the best diesel engine.

Salomon Sulzer-Bernet (1751–1807).

The Sulzer diesel engine has its origins in the late years of the 19th century. At the time, Sulzer Brothers was a substantial engineering enterprise, with a wide range of products, including iron castings, heating radiators, steam boilers, stationary steam engines, centrifugal pumps, ventilation fans, paddle steamers, and refrigeration machinery. The firm had been founded in 1834 by Johann Jakob Sulzer-Neuffert and his two sons, Johann Jakob and Salomon. Johann Jakob Sulzer-Neuffert had taken over the brass foundry and workshops of his father, Salomon Sulzer-Bernet in 1805.

On New Year's Day, 1834, the family decided to build a foundry, for which the foundation stone was laid on 7 April, on a new site by the road to Zürich. Father and sons applied their efforts to expand the business, especially by supplying castings to the textile industry, which was then developing rapidly in Switzerland. From the earliest times, they also made efforts to improve the standard of production, introduce new techniques, and to add new products. For the first eighty years, Sulzer Brothers was constituted as a partnership, principally with members of the family as partners. In June 1914, the firm was re-organised into a limited liability, joint stock company.

The remarkable association of the firm with the diesel engine stems from 1879 when, at the age of 21, Rudolf Christian Karl Diesel came to the firm's Winterthur works for some workshop experience after studying at the Technische Hochschule München. In the summer of 1879 he had suffered typhoid fever and missed his final examinations. Professor Carl von Linde of the Hochschule, however, arranged for Diesel to enter the Sulzer Brothers' works as a trainee to make good use of his time while waiting to retake his examinations in January 1880. Sulzer Brothers were then building ice-making machines

J.J. Sulzer-Neuffert (1782–1853).

for Linde, an authority on heat engines and refrigeration. Subsequently, Diesel gave a brilliant performance in his final examinations with the best grades since the Hochschule had been founded in 1868. He returned to Winterthur until, in March 1880, he was appointed to the firm in Paris that had the French sales rights for Linde's refrigeration machinery. He became the manager in Paris in January 1881 at the age of 23.

Birth of the Diesel engine

In the past, there has been much heated controversy over attributing the invention of the compression-ignition oil engine to Rudolf Diesel (in fact, toward the end of his life, Diesel suffered the outrage of colleagues, and virulent criticism of his integrity from many quarters of the German scientific community). Despite the many advances in the development of practical internal-combustion engines made in the 19th century, however, only Diesel's work has stood the test of time in large engines.

Diesel's novel achievement was to arrive at his engine concept by thermodynamic reasoning, quite unlike contemporary engine inventors of the late 19th century who relied on trial and error. Thus, he effectively foreshadowed modern design practice in which detailed theoretical analysis is combined with comprehensive experimental programmes. Just as Diesel stood astride two centuries, he lived through and contributed to the birth of a new profession based on scientific reasoning: engineering.

Diesel applied for his landmark patent (DRP 67207) on 28 February 1892, and it was granted one year later. As set out in his book, *Theorie und Konstruktion eines rationellen Wärmemotors zum Ersatz der Dampfmaschinen und der heute bekannten Verbrennungsmotoren (Theory and design of an efficient heat engine as the replacement for steam engines and today's known combustion engines)* published in January 1893, Diesel envisaged a prime mover of unprecedented efficiency.

His original concept was to employ Carnot's isothermal expansion cycle, in which high efficiency could be achieved by increasing the temperature difference in the engine cycle. He thus envisaged maximum combustion pressures up to 250 atmospheres, which was not realistic, so he came down to a more practicable level of about 35 atmospheres in engine tests. Nevertheless, this was still considerably higher than the 10 to 15 atmospheres used in contemporary oil engines. He also had to discard the quite impracticable Carnot cycle in

The original Sulzer Brothers: Salomon (1809–1869) and Johann Jakob (1806–1883).

Rudolf Christian Karl Diesel (1858–1913), in a triumphant moment before tragedy.

favour of one with constant-pressure combustion. That Diesel had to later revise his concept to develop a realistic engine does not detract from the fundamental importance of his work in creating a true compression-ignition engine, achieving a dramatic improvement in thermal efficiency, and of his advanced use of analytical techniques in the development of his engine.

During 1892, Rudolf Diesel sought backing for the development of his engine. He wrote to Maschinenfabrik Augsburg (which become MAN in 1908), Gasmotoren-Fabrik Deutz, and Mannesmann-Werke of Berlin, but without success. Later, Maschinenfabrik Augsburg wrote again expressing their willingness to build an experimental engine; an agreement for this purpose was signed on 21 February 1893.

Diesel also attempted to promote his idea elsewhere. In January 1893, he sent copies of his new book to various professors and heads of other major engineering firms, including Sulzer Brothers in Winterthur, Körting in Hannover, and Friedrich Krupp in Essen. Krupp later joined Maschinenfabrik Augsburg to help share in the development costs for the new engine.

PATENT-URKUNDE

№ 67207

AUF GRUND DER ANGEHEFTETEN BESCHREIBUNG UND ZEICHNUNG IST DURCH BESCHLUSS DES KAISERLICHEN PATENTAMTES

an Rudolf Diesel, Ingenieur, in Berlin

EIN PATENT ERTHEILT WORDEN.

GEGENSTAND DES PATENTES IST:

GESETZ v. 7. APRIL 1891

Arbeitsverfahren und Ausführungsart für Verbrennungskraftmaschinen.

ANFANG DES PATENTES: 28. Februar 1892.

DIE RECHTE UND PFLICHTEN DES PATENTINHABERS SIND DURCH DAS PATENTGESETZ VOM 7. APRIL 1891 (REICHS-GESETZBLATT FÜR 1891 SEITE 79) BESTIMMT.

ZU URKUND DER ERTHEILUNG DES PATENTES IST DIESE AUSFERTIGUNG ERFOLGT.

Berlin, den 23. Februar 1893.

KAISERLICHES PATENTAMT.

Beglaubigt durch Frank

Bureau-Vorsteher des Kaiserlichen Patentamtes.

Rudolf Diesel's German engine patent, granted 28 February 1892.

On 2 February 1893, Wilhelm Züblin (1846–1931), then chief engineer at Sulzer Brothers, visited Rudolf Diesel in Berlin to discuss his book. Among others, Züblin questioned whether an engine built to Diesel's theory could produce useful power. He argued that, since work input during the compression stroke was critical, relative to the power an engine produced, it "may happen that the loss is as great as the output."

Diesel followed up the meeting with a letter to Züblin in Winterthur on 8 February enclosing a 12-page paper outlining his expectations for the engine's performance. In it, he explained the advantages of his engine over steam engines, expressing great confidence that his method would prove beneficial, and intimating that a "very active business connection" was possible. In the accompanying calculations, Diesel showed that a thermal efficiency of 55 per cent was theoretically possible, using a compression pressure of 89 atmospheres and compression temperature of 800° C. He forecast that up to 47 per cent efficiency was practical, but he thought that 37 per cent was more realistic. Even the latter, lower value was a dramatic improvement compared with existing steam engines.

Correspondence continued but, at Sulzer Brothers, opinion was divided between those who were convinced of the future of Diesel's prime mover and those who shook their sceptical heads at launching a product based upon unproven technology. Since the 1850s, steam engines had been an important product for Sulzer Brothers and had gained the firm international recognition. (They had, for example, built two 6,000-horsepower (hp) compound engines, each with a single high-pressure cylinder flanked by two low-pressure cylinders, weighing more than 450 tonnes.) It was clear that by the turn of the century the

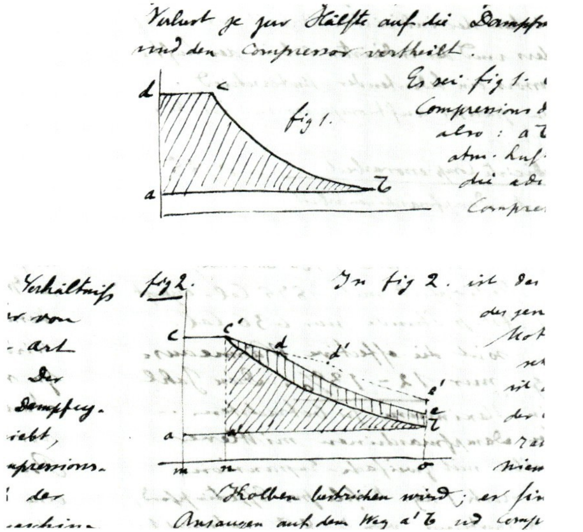

Diesel's sketches for his engine cycle.

...brüder Sulzer
Winterthur.

...e in Ludwigshafen a. Rh.

16. Mai 1893

VERTRAG

zwischen den Herren Gebrüder Sulzer in Winterthur einerseits,
und Herrn Rudolf Diesel, Ingenieur in Berlin, anderseits.

§. 1.

Herr Diesel ist Besitzer des schweizerischen (provisorischen) Patentes No.5321, betitelt: "neuer Verbrennungsmotor" und überlässt den Herren Gebrüder Sulzer das Recht der Ausführung und des Verkaufs der nach diesem Patent arbeitenden Motoren, und zwar für die ganze Schweiz derart, dass genanntes Recht in der Schweiz vom jeweiligen Patentinhaber weder ausgeübt noch anderweitig übertragen werden darf. Die Ausführung darf nur in eigener [illegible] ...egenden [illegible] gegen Uebernahme de... kosten ohne Weiteres zur Benutzung zu überlassen.

Also übereingekommen, doppelt ausgefertigt, beidseitig unterzeichnet und ausgewechselt.

Winterthur und Berlin, den 16. Mai 1893.

Diesel.

Letterhead and signature pages for the agreement between Diesel and Sulzer Brothers.

Sulzer works was well practised in building large reciprocating prime movers, long before diesel engines of comparable size were built.

Even those in favour of the new engine concept, led by Johann Jakob Sulzer-Imhoof, were not prepared for the time being to venture beyond a purely scientific interest. The firm's eventual decision was therefore to bide their time while closely watching developments. Eventually, on 16 May 1893, the firm signed an agreement with Rudolf Diesel to secure the exclusive use of his patents in Switzerland together with an option to take up engine manufacture later. The firm paid him 10,000 Marks initially, followed by an annual sum of 20,000 Marks half-yearly until they decided to start manufacturing engines themselves. The agreement also gave Sulzer Brothers the right to receive all results and improvements arising from research at Maschinenfabrik Augsburg and Krupp. Rudolf Diesel was also obliged to let Sulzer Brothers have whatever research results and design drawings the firm wished to have from Augsburg and Krupp.

This was one of the original license agreements granted by Rudolf Diesel before the holding company Allgemeine Gesellschaft für Dieselmotoren AG (Allgemeine) was set up in 1898 to handle all rights to Diesel's patents. Beginning in 1893 Rudolf Diesel signed license agreements with ten companies in Switzerland, Germany, France, Belgium, Scotland, Denmark, and the USA. Diesel's first engine was completed in Augsburg in July 1893, but did not run independently until February 1894. It was a vertical, single-cylinder engine of 150-mm bore and 400-mm stroke, with an "A"-frame base, crosshead, and a long plunger-type piston. It had a compression pressure of 30 atmospheres, with a compression temperature of about 500° C. It took nearly three years more for Diesel and the Augsburg engineers to bring the engine to an operational condition. In this time the test engine was rebuilt twice, and its bore was increased to 220 mm. It used liquid fuels from the outset (coal was tried for a short time in 1899), but it was changed from airless to air-blast fuel injection and subsequently fitted with an integral injection air compressor. A new engine was completed in October 1896. The bore was increased to 250 mm.

Johann Jakob Sulzer-Imhoof
The principal advocate of the diesel engine at Sulzer Brothers

Johann Jakob Sulzer-Imhoof (1855–1922).

was Johann Jakob Sulzer-Imhoof. His confidence in the capacity of this type of engine to develop high powers led to the introduction of the two-stroke reversing diesel engine for ship propulsion, and to the development of diesel engines for railway traction. Elder son of Salomon Sulzer, one of the founders of the firm in 1834, he was an active partner in the firm from 1889 to 1914, when it became a limited company, and then its chairman until he retired in 1920. Before joining the firm in 1883, Sulzer-Imhoof had studied at the Federal Polytechnikum (now the ETH) in Zürich and at the Dresden Technische Hochschule, and widened his experience and practical knowledge by working on steam engines and locomotives at Carels Frères at Ghent, and in shipbuilding at Lobnitz & Co in Renfrew and R. Napier & Sons in Glasgow. Sulzer-Imhoof was regarded as a born engineer, his chief personal attribute being the will to tackle the most difficult mechanical problems in a most courageous way.

Sulzer-Imhoof is known to have visited Augsburg in November 1895 to see Diesel's first engine. He then returned in 1897 to see the second engine being tested. During the first weeks of 1897, the engine was demonstrated to visitors from various firms, including Diesel's licensees. On 17 February 1897 Professor Moritz Schröter completed a series of independent, official tests on this new second engine. At full load, it developed 17.8 horsepower at 154 rpm, with a specific fuel consumption of 238 grams per horsepower-hour at full load. This was equivalent to a thermal efficiency of 26.2 per cent. In later tests, the thermal efficiency could be increased to 30.2 per cent.

This successful demonstration quickly attracted attention and Diesel's presentation of his engine to the VDI meeting on 16 June 1897, supported by Professor Schröter's report on the tests, was triumphant. At a time when steam engines had efficiencies in the order of 10 to 15 per cent and hot-bulb oil engines were only a little better, this quantum jump in engine performance was a remarkable achievement.

First Sulzer diesel engine

After the successful running of the new test engine in Augsburg, the various Diesel licensees began to build their own research engines, using drawings from Rudolf Diesel's design office in Augsburg.

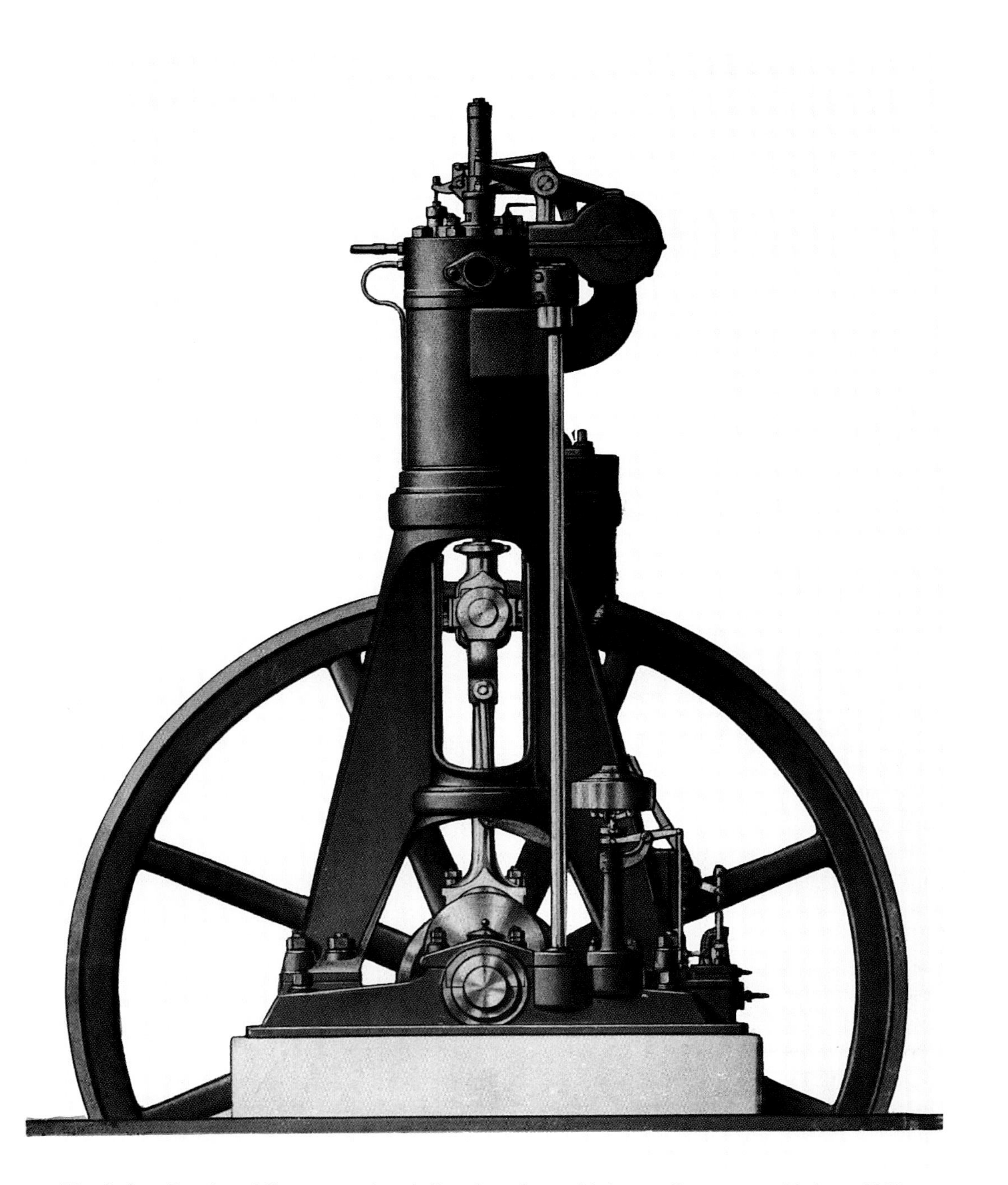

The Sulzer Brothers' first operational diesel engine, which was first run on 10 June 1898.

As one these licensees, Sulzer Brothers began the manufacture of a 20-horsepower single-cylinder test engine on 11 August 1897. It was a four-stroke engine of 260-mm bore and of similar crosshead-type configuration to Diesel's engine in Augsburg.

Erection of this engine was completed in May 1898. It was set up to drive an adjacent line shaft from which it could also be driven. The engine was loaded by a centrifugal pump driven from the line shaft. The engine was first run on 10 June 1898, a prime day in technological history. The first indicator diagram was taken on 21 June.

After a difficult period of running-in and many modifications, Sulzer Brothers sent a report on the first tests to the Allgemeine. According to this report, the engine ran without overheating and the bearings remained completely cool, but its mechanical efficiency was only 53 per cent. However, by the late summer 1899, the engine's performance had been improved. The licensee-built research engines marked the beginning of a lengthy period of intensive research, experimentation, and design improvement before the diesel engine concept became a commercial reality.

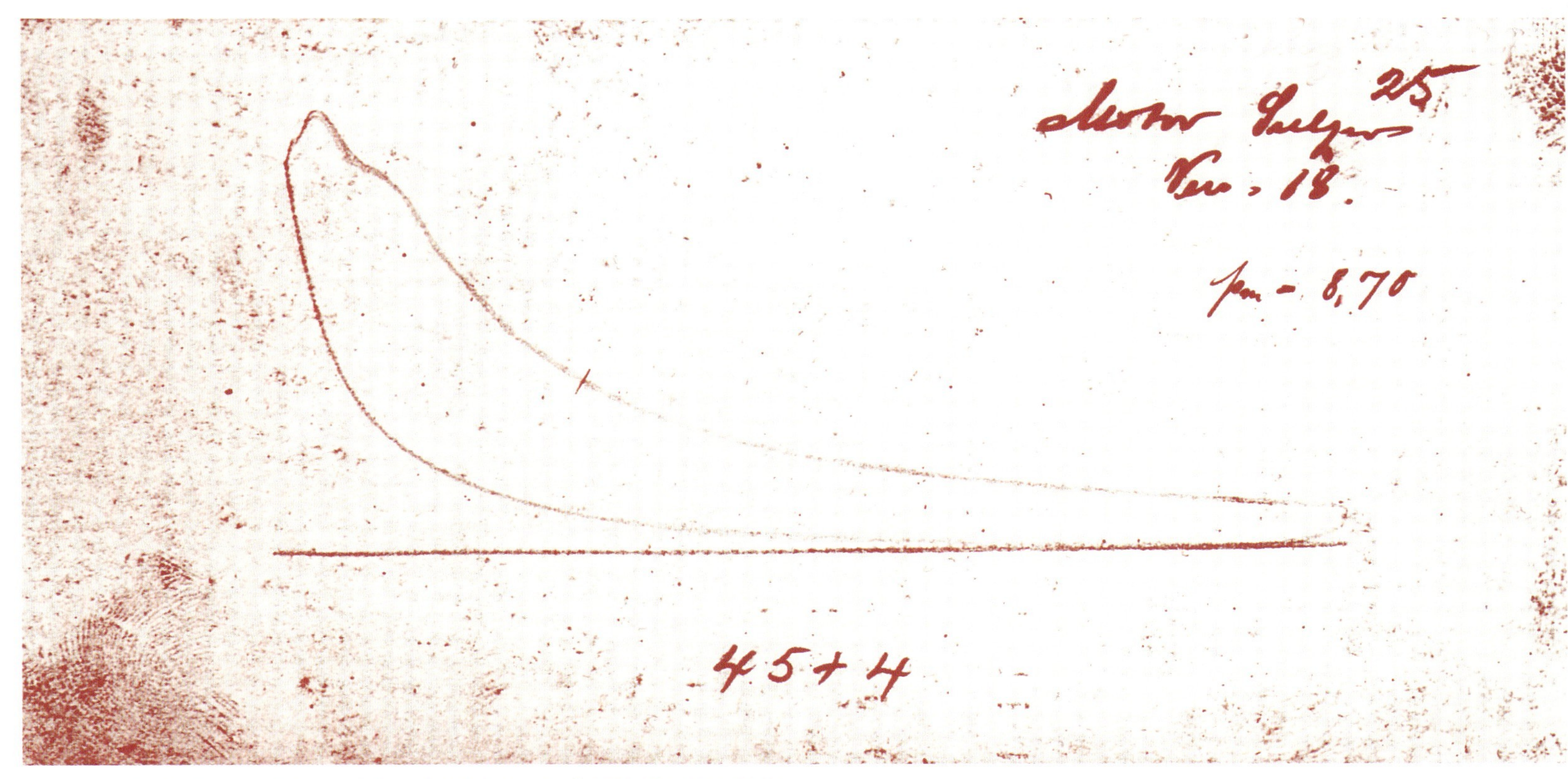

An early indicator diagram taken from the first Sulzer diesel engine.

Sulzer Brothers' cautious approach to diesel engine development proved to be well justified. By the end of 1898, a number of production engines supplied by Maschinenfabrik Augsburg and other licensees, soon proved premature: Their initial operating results gave cause for alarm and shook the confidence of even the most confirmed diesel enthusiasts. As a result, regular engine manufacture in Winterthur was delayed while many "teething" problems were overcome.

The watershed for the emerging diesel technology came with the 1900 Paris World Fair, at which an 80-hp, two-cylinder diesel engine built by the French licensee won the power division's Grand Prix. By May 1902, some 359 diesel engines totalling 12,367 hp were in service or on order. The diesel engine was quickly being applied worldwide.

Manufacture in Winterthur

Sulzer Brothers delayed starting the manufacture of diesel engines because the 1893 agreement did not afford them adequate market outlets outside Switzerland to make it a viable business proposition. However, on 23 April 1903 the firm signed a new license agreement with the Allgemeine, for the manufacture of diesel engines which, apart from giving Sulzer exclusive rights in Switzerland,

One of the twelve Sulzer 1D40 35-hp engines built in 1903.

also enabled them to export worldwide. Apart from export opportunities through the Diesel Engine Co Ltd (which had been set up in London by the Allgemeine to sell engines in the British Empire), Sulzer could also sell engines in any other country, providing it reached an agreement with the respective patent holder.

The agreement was made for a duration of 12 years. A premium of 60,000 Marks was to be paid when the agreement came into force. Royalty fees were set basically at a percentage of the engine selling price, but were reduced for engines exported to those countries where Sulzer had reached an agreement with existing licensees. No royalties were due on engines delivered to countries for which the Diesel Engine Co held the patent rights.

The Diesel Engine Co had already placed a pilot order with Sulzer for twelve 1D40 single-cylinder engines each of 35-hp output in January 1903. It was seen as essential for the firm's reputation that the new engines be successful. So Sulzer made a prototype to remedy problems and test innovations. The 1D40 was running by the end of June 1903, when the first of the production engines was to be shipped.

In early 1903, Sulzer Brothers established a diesel engine department under Johann Jakob Sulzer-Imhoof, assisted by Fritz Schübeler (1849–1927), who had worked with him on the building of passenger steam vessels, primarily for the Swiss lakes. Within the department, Sulzer-Imhoof set up a design department for four-stroke engines, a research office for two-stroke engines, and a sales department. From the outset, Sulzer Brothers set out to rationalise engine production by offering a standard series of engines with one or more cylinders to give a broad power range for stationary applications.

The D-type engines were vertical trunk-piston models of heavy, open A-frame construction in cast iron, with a high-level camshaft. Fuel oil was blown into the cylinder by compressed air (blast injection), provided in the early engines by a reciprocating compressor driven by a rocking lever from the piston. They operated at modest ratings with about five bar mean effective pressure and three to four metres/second mean piston speed, with a compression pressure of 30 to 35 bar. The D-type engine proved suitable for a wide range of stationary duties, including driving pumps, generators, refrigeration compressors, factory line shafts, and other industrial machinery. Engines quickly became geographically widespread, as far afield as Great Britain, Egypt, New Zealand, and South Africa.

There eventually came a need for lighter, faster, more compact engines. Accordingly, between 1910 and 1913, Sulzer introduced the K-type four-stroke series, with shorter strokes than the D-type. They used box-column construction in which individual cylinder jackets were mounted on a common crankcase, and had similar bore sizes to the D-type. Together, the D and K engines initiated a line of Sulzer four-stroke engines, culminating in the advanced designs of today's ZA40S, ZA50S, and S20U.

At about the same time as the K-type engine was becoming established, Rudolf Diesel, in 1913, published an exposition on the origin of his engine: *Die Entstehung des Dieselmotors* (1913). He had been in poor health, and after earning a fortune from his license fees and royalties in the early years of the new century, he had serious financial setbacks. He built a mansion for himself and his family in Munich that he could no longer afford; he was facing bankruptcy. His early patents had expired, and he was constantly subjected to public vilification from a few prominent members of the scientific community who accused him of a lack of originality in his invention. He told family members that he was having heart problems, experiencing shooting pains in his head, and was reading the works of Schopenhauer.

Nonetheless, J.J. Sulzer-Imhoof and Rudolf Diesel continued their genuine friendship. Clearly, Diesel had a high regard for Sulzer-Imhoof, as witnessed by the inscription he wrote in the copy of his exposition he gave to his friend: "Dedicated to Mr Sulzer-Imhoof in friendship for our years of work together. Munich, September 1st, 1913. [Signed] Diesel." Twenty-eight days after signing that inscription, Rudolf Diesel and two Belgian colleagues boarded the steamer *Dresden* in Antwerp, to cross the Channel to Harwich, England. The next morning Rudolf Diesel was nowhere to be found aboard the vessel. In early October a Dutch pilot vessel hauled a body from the sea, removed all personal effects and, as was the custom, committed the body back to the sea. The effects belonged to Rudolf Diesel.

Herrn Sulzer-Imhoof Zum Andenken an unser jahrelanges gemeinsames arbeiten in Freundschaft gewidmet.
München, den 1. September 1913.
Diesel.

DIE ENTSTEHUNG
DES DIESELMOTORS

VON

RUDOLF DIESEL
DR. ING. H. C.
DER TECHNISCHEN HOCHSCHULE MÜNCHEN

MIT 83 TEXTFIGUREN UND 3 TAFELN

BERLIN
VERLAG VON JULIUS SPRINGER
1913

The copy of his final publication, which Rudolf Diesel inscribed to J.J. Sulzer-Imhoof.

Increased capacity
The increased demand for Sulzer diesel engines led, in 1907, to their manufacture also being established in the firm's Ludwigshafen works in Germany which, until then, had mainly built steam engines. In July 1909, Sulzer Brothers also granted a license to Schweizerische Lokomotiv- und Maschinenfabrik (SLM), next door in Winterthur, to manufacture small- and medium-sized four-stroke diesel engines to the firm's designs.

This pioneering move by Sulzer Brothers to license the manufacture of diesel engines in the early years was a stunning milestone in the history of technology. It led soon to the first licensing in Japan, and eventually to the structuring of the entire diesel engine industry as a series of partnerships between designers and manufacturers that still obtains today.

The increase in production and the promise of increased demand soon mandated the extension and modernisation of the firm's Winterthur facilities. In 1909/1910 Sulzer Brothers demolished the old buildings and replaced them with new workshops with large erection halls and test beds. The shops were equipped with up-to-date machine tools and cranes.

The old test beds had been relatively small and narrow. Engines were erected in one place and then dismantled and erected again on a test bed. In the new workshops and test beds, the same procedure was continued for the larger engines but, with the new cranes, small and medium-sized engines could be transferred directly to the test beds without dismantling.

This was taken a step further in a new (1931) erection hall, where even the largest engines could be tested where they were erected. The dynamometers instead were moved and the shop was well served by connections for engine services — cooling water, lubricating oil, compressed air, electricity, and exhaust outlets.

The early engines were usually tested by their driving a dynamo through belting. Some of the faster-running engines, however, were also tested using friction-band absorption brakes. Then, between 1907 and 1910, the Sulzers bought a few Froude water dynamometers and installed them in the new diesel workshops. These dynamometers gave much better results in testing and Sulzer soon added more of them to be used for all future engine acceptance tests.

End of Diesel's license
After 1903, there were various quarrels among the Allgemeine licensees about the mutual obligation in the agreements to freely exchange technical information, design improvements, engine drawings, and other experience. Although an important feature of the agreements, it proved to be a major weakness and source of tension. All participants were afraid of giving more than they were receiving.

In Sulzer Brothers' case, after engine manufacture had been started in 1903, the firm put considerable effort into engine development, producing new engine designs and developing practical two-stroke engines. According to the license agreement, these new designs were supposed to be given freely to other licensees. On the other hand, Sulzer had the right also to export engines worldwide, in competition with the same fellow licensees with whom they were expected to exchange drawings.

Eventually in 1907, Sulzer Brothers sued the Allgemeine to release the firm from the obligation to give drawings and other technical information to the Allgemeine and other licensees. The Allgemeine made a counterclaim that the payment of royalties had to continue. The suit regarding information transfer was won by Sulzer, but they lost on the matter of their royalties. Differences were finally settled by an agreement in January 1909, which annulled all existing contracts between the Allgemeine and Sulzer Brothers, and cancelled all remaining patents. At the same time, royalty payments were to be stopped.

In return, Sulzer Brothers renounced a 5,000-Mark credit with the Allgemeine and various amounts owing from the Allgemeine's French and Italian licensees. Sulzer Brothers also had to pay all royalties owing to the Allgemeine under the 1903 contract. There was, however, little purpose for the Allgemeine to exist after the expiration of the original patents in Germany, and it was dissolved in 1910.

It has been estimated that from 1893 to 1909, Sulzer Brothers paid about 350,000 German Marks for the Swiss rights to Rudolf Diesel's patents. In modern values, this is probably equivalent to some five million Swiss Francs, which is a considerable amount to stake on a new product whose technology was unproven and whose commercial future had not been very clear.

The Sulzer Areal, about 1931, with the engine assembly hall *(centre)*. A modern office complex was later built in the foreground triangle.

First marine applications
Although the earliest diesel engines were used for power generation, their advantages of low fuel consumption, compact installation, safe operation, and ready cold-starting made them attractive for marine propulsion. For this, however, it was also necessary to develop the engine's ability to vary speed, and change its direction of rotation for the ships manoeuvring and reversing.

The French canal barge *Petit Pierre,* in September 1903, became the first vessel with a diesel engine. She had a 25-hp, single-cylinder horizontal engine built by Sautter-Harlé and coupled to a reversing propeller. Then followed a series of Nobel tankers on the Volga River and Caspian Sea. They were powered by engines supplied by AB Diesels Motorer, Stockholm, and Ludwig Nobel Ltd of St Petersburg. With these engines, reversing and speed variation were provided by electric transmissions.

Venoge, a Lake of Geneva cargo vessel (1905), and one of the earliest diesel-driven vessels.

The first vessel equipped with a Sulzer diesel engine was a canal boat of Leeds-Liverpool Canal Co in Great Britain. The vessel was outfitted with a 30-hp twin-cylinder 2D15 four-stroke engine installed in May 1905. Unfortunately, it was not a success, probably because of its poor manoeuvrability. It was a standard stationary engine with unidirectional rotation and it used either a reversing propeller or a reversing gearbox. The engine was later transferred to the canal company's yard to drive a saw mill, where it worked to at least 1922.

The 45-horsepower twin-cylinder Sulzer D20 engine, which entered service in the cargo vessel *Venoge* on the Lake of Geneva in September 1905, was far more successful. It was also a standard D-type four-stroke engine. Although it was non-reversing, the engine drove the vessel through a Del Proposto system in which ahead propulsion at full power was by accomplished by direct transmission to the propeller, with astern drive and variable-speed manoeuvring in both ahead and astern directions being given by a DC electric transmission.

Venoge's 45-hp, two-cylinder D20.

The Milan reversing marine engine, which won a prize, but was never installed in a ship.

Early two-stroke engines

From the beginning, Sulzer Brothers had envisaged the development of large diesel engines, particularly for ship propulsion, and thus recognised the special advantages of the two-stroke cycle. Even when manufacture began in Winterthur, diesel engines all operated on the four-stroke cycle with natural aspiration and were very limited in output. But, by creating simpler engines with nearly double the cylinder output, the two-stroke cycle was seen to have a better chance of meeting the powers needed for ocean-going ships. Therefore, J.J. Sulzer-Imhoof established a research office under Arnold Lack to study two-stroke engines and the application of two-stroke engines in ship propulsion and rail traction. In 1904 Walter Schenker began a systematic study of the scavenging process.

Two variants of two-stroke scavenging were investigated. Port scavenging was more attractive because it was structurally simple. The cylinder had inlet and exhaust ports, controlled by the piston. The cylinder head needed to accommodate only the fuel and starting valves. But, with exhaust ports higher than the air ports, the cylinder could never be charged to the full scavenge air pressure. Valve scavenging, on the other hand, provided for the scavenge air to enter through valves in the cylinder head, and exhaust gases to leave through ports around the lower part of the cylinder, which were uncovered by the piston. The cylinder could thus be charged to the full scavenge pressure by simply holding the scavenge valves open for a brief spell after the exhaust ports were covered.

In both variants, the problem was to obtain the highest possible purity of air charge in the cylinder, with the least exhaust gases remaining, while using the least amount of excess scavenge air and the lowest scavenge pressure. Tests in 1905 using a full-size cylinder showed that valve scavenging was more efficient than port scavenging. Valve scavenging was thus employed in the first Sulzer direct-reversing two-stroke diesel engine ordered in 1905. The engine was exhibited at the Milan World Fair in 1906. This four-cylinder version, the DM175/250, which developed 90 horsepower at 375 rpm, attracted considerable interest and won a Grand Prix.

The Milan engine was never installed in a vessel, but was the basis of a new series of two-stroke engines, the S-type, announced in 1907, of which there were ten models with outputs of 64 to 1,600 hp. They had twin overhead camshafts directly operating the two scavenge-air valves in each cylinder. Scavenge air was

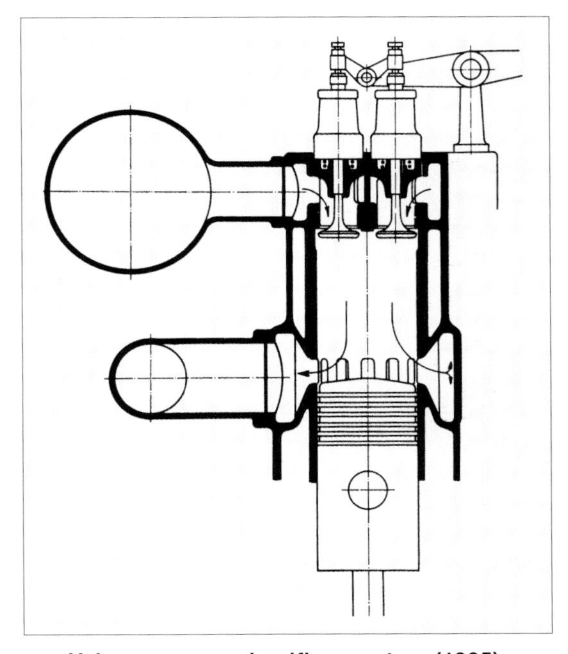

Valve-scavenged uniflow system (1905).

provided by a double-acting pump at the free end of the engine. As in all diesel engines up to about 1930, fuel was injected by compressed air (blast injection). Starting, reversing, and manoeuvring were from a single control wheel. The first of the new series was installed in an Italian naval tug in 1909. A slightly larger engine was built in 1909 for the 1864 Lake of Zürich steamer *Taube*.

Simple port scavenging

Valve scavenging proved to be just an interim measure. In 1909, Sulzer changed to cross-flow port scavenging for marine and stationary two-stroke engines; without the valves, cylinder cover design was simplified. Two rows of scavenge air ports faced the exhaust ports. The lower scavenge ports were controlled by the piston, while the upper ones were controlled by cam-operated double-beat drop valves, which prevented exhaust gases blowing back into the air manifold as the upper ports were uncovered by the piston. After scavenging was completed, the cylinder was charged up to the scavenging pressure through the upper ports. This led to engines with a modest power increase, lower fuel consumption, and greater reliability. Further successful developments followed.

Walter Schenker can truly be regarded as the father of the Sulzer two-stroke engine. He undertook the early research that enabled port scavenging to work and thereby made possible the simple, valveless engine. He began the search for the best port forms by cleverly adapting a cigar box with a glass cover to form a flat model of a cylinder into which he could insert various shapes of scavenge ports. By blowing cigar smoke through the ports, he could visualise the scavenge air flow.

The first Sulzer electricity-generating two-stroke engine, installed in Aarau, in 1908.

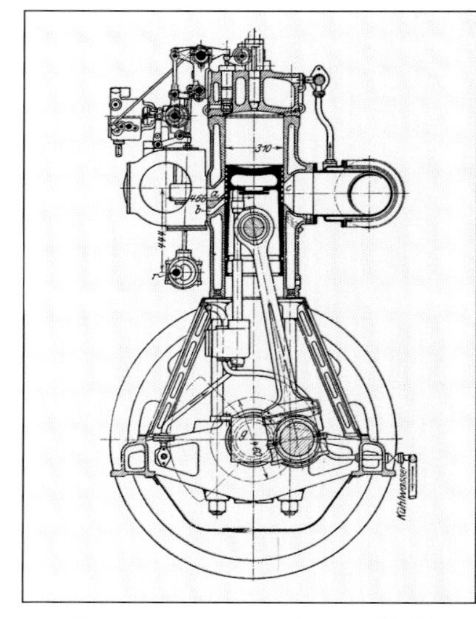

Port-scavenged S-type (1909).

Schenker joined the Sulzer diesel engine department in April 1903, after experience in steam engine design at Sulzer, and gas engines at Körting and Tosi. Evidently well regarded as an engineer, he had an inventive talent and initiated very many technical solutions. Schenker became technical director for the diesel engine department in 1920, and retired in 1935.

Schenker was one of a long line of notable chief engineers responsible for all technical matters concerning Sulzer diesel engines.

More marine applications

The first two port-scavenged marine engines, of 380 horsepower each, were ordered from Sulzer in 1909, and installed in the 1,000-dwt Italian passenger/cargo vessel *Romagna,* which entered service in October 1910. After early teething problems with various aspects of the machinery installation, the vessel was reported to be entirely satisfactory once in service. The engines met the high reliability needed by the regular service along the Adriatic coast and gave the expected economic savings. They were noted to be just as manoeuvrable as steam engines. Unfortunately, the *Romagna* capsized and sank in November 1911, when badly stowed cargo shifted during a sudden storm. The loss, however, was greater than that of a unique ship alone; all hands went down with her as well, including a Sulzer engineer who was doing his duty to the owner to see that the new engine was operating as promised.

By 1912, a total of 365 ships with diesel engines from several manufacturers were thought to be in service, among them the East Asiatic Co's *Selandia*, powered by twin 1,010-hp Burmeister & Wain

The Italian passenger/cargo vessel *Romagna* (launched, but later doomed).

Romagna was powered by twin 380-hp 4SNo.6a direct-reversing engines.

four-stroke engines, launched that year. In August 1912, the 6,324-dwt Hamburg-Süd cargo liner *Monte Penedo* entered service with twin Sulzer 4SNo.9a two-stroke engines, of 470-mm bore with a combined output of 1,700 hp. They were also the first marine two-stroke crosshead engines from Sulzer. In fact, she was the first large ocean-going vessel to be powered by two-stroke diesel engines and had a long life. She was scrapped in 1969.

Metre-bore research engine

In the same month that the *Monte Penedo* entered service, Sulzer began testing a one-metre-bore, single-cylinder experimental two-stroke engine, the 1S100, which was used until mid 1915. The engine was a cross-scavenged, crosshead-type, using double scavenge ports. Built to explore the possibilities of very large two-stroke engines, the 1S100 answered many other questions concerning combustion, fuel

Monte Penedo, **a 6,324-dwt cargo liner, entered service in August 1912. She was powered by twin Sulzer 4SNo.9a two-stroke engines *(top).***

valves, lubrication, bearing performance, scavenging efficiency, and thermal strains in the cylinder head, cylinder liner, and piston. Sulzer technicians also saw the engine as a possible contender for warship propulsion, for which its overall height of less than six metres was an important feature.

Between 1911 and 1920, the stationary market was the most important sales application for Sulzer two-stroke engines. The largest Sulzer diesel engine of that period was the 6Z300 stationary engine ordered by Harland & Wolff in 1912 for their Belfast shipyard. With six cylinders of 760-mm bore, this crosshead-type port-scavenged engine developed 3,750 hp at 132 rpm.

All together, 163 stationary Z-type engines, with valve or port scavenging, blast or airless injection, single- or double-acting, were built over the years with a combined output of some 375,000 horsepower. They included some notable power generating installations in Calais, Shanghai, Bucharest, and Broken Hill, Australia. This last plant was closed in 1986, after 55 years' continuous service, and its engines had reached a total of 2,450,213 running hours.

The metre-bore research engine of 1912.

Across the Seven Seas

Though only small engines were sold during World War I, the adoption of diesel engines for ship propulsion really widened after the war with the need for fleet replacements. The first notable units were two four-cylinder S60 engines that were ordered for stock in September 1917. They were sold in 1919 and installed in the 6,500-dwt Spanish tanker *Conde de Churruca* delivered by the Armstrong Whitworth

Lokawei, Shanghai, for a time the most powerful diesel power station in the world.

yard in 1921. Between 1908 and 1931, Sulzer and their licensees built a total of 604 S-type blast-injection, two-stroke marine engines with an accumulated output of 1,132,110 horsepower. They were built in Winterthur, Ludwigshafen, and Cie de Construction Mécanique Procédés (CCM), and at licensees in Great Britain, Italy, Japan, the Netherlands, Germany, the USSR, and Spain.

The world's first large passenger liner propelled by diesel engines was the 17,490-grt *Aorangi*, built at Fairfields for the Union Steamship Co of New Zealand in December 1924. Her quadruple-screw machinery comprised four six-cylinder Fairfield-built ST70 two-stroke single-acting engines of 700-mm bore, aggregating 13,000 hp at 127 rpm. The positive service experience obtained with these engines, and in other contemporary motor ships, helped to dispel the remaining prejudices against the use of diesel propulsion in large ships.

Among other ships built about the same time, to be equipped with Winterthur engines, was the French 8,898-grt passenger/cargo liner *Camranh,* with twin 4ST68 engines totalling 4,000 hp; Japan's first large motorship, the 10,040-dwt NYK cargo vessel *Atago Maru* built on the Clyde with twin 4ST68 engines; and seven *Bel*-class heavy-lift ships built between 1924 and 1928 by Armstrong Whitworth, the first ships able to load heavy cargoes, such as fully-assembled railway locomotives, using their own gear.

In 1929 and 1930 NYK took delivery of two 16,975-grt quadruple-screw liners, *Asama Maru* and *Tatsuta Maru,* each driven by four 8ST68 engines, together giving 16,000 hp, half of the engines being made by Mitsubishi in Japan. The Italian *Victoria* and *Neptunia* (1931 and 1932) quadruple-screw

Conde de Churruca **(1924), with the first large two-stroke engines after World War I.**

The liner *Aorangi* (1924), a rare quadruple-screw vessel with 13,000 total horsepower.

liners had four CRDA-built 8ST68 engines. (The 13,062-grt *Victoria* was briefly the world's fastest motor ship, at 23.26 knots.)

One of the leading countries in changing over to large diesel engines for deep-sea shipping was the Netherlands. Diesel engines were well suited to their long-haul liner services to the Dutch East Indies (now Indonesia), where there were ample supplies of good-quality fuel oil. The Dutch initially converted two cargo steamers, *Wieringen* and *Bintang,* from steam to diesel propulsion, with 4ST60 and 6S76 engines. These steamers opened the way for a long line of Dutch ships that were powered by Sulzer diesel engines, including passenger mail liners and cargo ships, primarily for the Stoomvaart Maatschappij Nederland (SMN) and Rotterdam Lloyd.

In the 1930s, five Dutch vessels were refit with geared Sulzer two-stroke crosshead-type engines, driven through flexible couplings and twin-input/single-output gearboxes to the single propeller. Geared machinery was also employed for the eight direct-reversing two-stroke crosshead engines installed in the famous Royal Rotterdam Lloyd liner *Willem Ruys* (which, with 32,500 horsepower, achieved 24.62 knots on her sea trials). She did mail service to the Dutch East Indies. (*Willem Ruys* was sold in 1964 and renamed *Achille Lauro*. She was refitted to carry migrants to Australia and New Zealand, and was transferred to cruising in 1972. She was lost by fire in December 1994, still equipped with her original engines. A service report in late 1990 gave the remarkable total running time of about 250,000 hours for each engine.)

***Belpamela* (c.1924), a Norwegian heavy-lift ship with a full deckload of railway carriages.**

Another major installation of Sulzer two-stroke engines was in the famous SMN liner, the 20,017-grt *Oranje* in 1939. Her triple-screw machinery of Winterthur-built engines, with a combined output of 37,500 hp, gave her a speed of 26.3 knots. (She also carried six 1,800-hp Sulzer diesel auxiliaries.) *Oranje* remained the world's most powerful motorship for many years to come. (After sustained service during World War II, the *Oranje* was operated on the East Indies run until she was sold in 1964 to become the *Angelina Lauro*. She was destroyed by fire in 1979 in the Caribbean.)

An 8ST68 destined for NYK's *Asama Maru*.

At the other end of the working spectrum, a number of Sulzer S-type engines were also installed in luxury private yachts. The largest, and most notable among these great ocean-going playthings was the 2,000-grt *Nourmahal* owned by Vincent Astor. Built by Krupp Germaniawerft in 1928, she had twin Winterthur-built 6S47 engines with a combined power of 3,200 horsepower.

The Dutch-built *Willem Ruys* with her engine control desk (1947).

Oranje, with a speed of 26.3 knots then the world's fastest merchant ship (1939).

Vincent Astor's classically handsome *Nourmahal*, with 3,200 horsepower of Sulzer engines.

The Battle of the Cycles

A tremendous, often acrimonious, rivalry arose in the 1920s and beyond, between the technical and professional advocates of the four-stroke cycle and those who favoured the two-stroke cycle for ship propulsion. The competition was often referred to by those in the industry, and by the marine press, as the "Battle of the Cycles."

Sulzer Brothers was one of the first international engine building firms to recognise the special advantages of the two-stroke cycle over the four-stroke cycle, and to concentrate their energies in engine development on the two-stroke engine for ship propulsion.

The four-stroke single-acting engine, however, had gained an early lead in the marine market. Initially, only a few other firms took up the two-stroke cycle along with Sulzer before World War I.

Sulzer's concentration on the two-stroke cycle, backed by systematic research and development in their Winterthur facilities, gave the firm an important technological lead on competitors who adopted the two-stroke cycle only later. By the end of the war, Sulzer had already created the valveless two-stroke engine for which it ultimately became recognised around the world.

At the same time, the firm had accumulated considerably more experience than its competitors with designing and building efficient and highly reliable two-stroke engines. By the middle of 1919, more than 300,000 hp of Sulzer two-stroke diesel engines had been delivered to clients or were in the course of construction.

As ships became larger and faster, the two-stroke cycle in both single-acting and double-acting forms became the predominant form of propulsion. The two-stroke cycle allowed higher power outputs through the then usual natural aspiration, while valveless cylinder covers would perform more reliably than the overly complicated designs for four-stroke engines, with their valves.

By the end of the 1920s, the battle had been largely won, and the builders of large, four-stroke marine engines gradually changed over to building two-stroke engines. As further proof that the battle had been won, other builders of two-stroke engines also entered the marine market. The two-stroke cycle reigned supreme for large, low-speed marine diesels. And, as is abundantly clear from the market, it still does today.

A cartoon of 1923 makes light of the serious war between two- and four-stroke advocates.

Double-acting engines

In the 1920s and 1930s, double-acting engines became a major line of development among marine engine builders as a means of increasing unit outputs. Studies into double-acting engines began at Sulzer in 1922, but it was only in 1927 that the DZ90 single-cylinder double-acting research engine became operational. Of 900-mm bore by 1400-mm stroke, the DZ90 had the largest bore of all double-acting diesel engines ever built. Generally, the DZ90 demonstrated that large-bore double-acters would be feasible.

A number of licensees embarked on the manufacture of double-acting engines, including Royal Schelde, in the Netherlands, and Kobe Steel and Mitsubishi Nagasaki, in Japan. The highest double-acting output was reached by two 760-mm-bore, 10-cylinder units of 14,000 hp built in 1934 under license by CRDA in Trieste, for refitting the Italian passenger liner *Saturnia*.

During test-bed trials in December 1935, one of these engines developed a record-breaking output of 20,820 horsepower at 152 rpm. The ship returned to service in August 1936 and she remained in operation until 1966.

Unfortunately, the double-acting engine had three intrinsic drawbacks: the piston rod was vulnerable to thermal cracking; poor combustion was inevitable in the lower cylinder; and the piston rod gland was difficult to keep sealed. In addition, the engines were very large and maintenance was both inconvenient and time consuming. Besides, Sulzer was never wholeheartedly in favour of that line of development. In the end, the Sulzer SD single-acting two-stroke marine engines in the 1930s virtually eliminated the weight advantage of the double-acting engines. With the SD types, it was possible to have a single-acting installation of virtually the same overall weight as one with double-acting engines. Together with the simpler construction of the single-acting engines, this meant that their price as installed was less than for a double-acting plant.

A very powerful 10DSDT76 engine of the type used in *Saturnia*.

The Italian liner *Saturnia*, refitted in 1936 with 28,000-hp double-acting Sulzer engines.

Naval applications

Throughout the time between 1910 and 1940, Sulzer Brothers developed a long line of compact two-stroke engines for submarines and other naval vessels, the Q types. The Sulzer Q-type engine proved to be particularly attractive for submarines. It was, in fact, the basis of many of the early license agreements: Before 1919, licenses for Q-type engines were granted to Ansaldo in Genoa; Ateliers & Chantiers de la Loire in St Denis, Paris; Kolomna Machine Works in Russia; the Norwegian navy dockyard; the Imperial Japanese Navy; and Suzuki & Co in Kobe, Japan. It was also the principal reason for the establishment in 1918 of the subsidiary Sulzer CCM, at St Denis, Paris.

The first Sulzer submarine engine was the 6U23; four were sold in 1910 and 1913 to the Italian Navy, two installed in the *Nautilus*, two in her sistership *Nereide*. The 6U23 was light for its day, achieved by the use of machined steel components where possible, instead of massive iron castings. The engine weighed about 25 kilograms per hp, yet nothing was sacrificed for safety, strength, and durability.

Until the 1930s, submarine engines were directly coupled to propellers, with electric motors for battery charging mounted on the same shafts. However, the long directly-coupled machinery system could cause torsional vibration problems. As a result, the US Navy pioneered the development of the all-electric drive for submarines, whereas European and Japanese navies continued to use the direct-drive arrangement throughout World War II. (Torsional vibration could generally be avoided by designing the drive train so that operating speeds did not coincide with vibration-inducing speeds.)

The US joint venture Busch-Sulzer Brothers-Diesel Engine Co entered the naval business in 1915 with two 600-hp 6U32 engines for the US Navy submarine *Turbot*. Subsequently, Busch-Sulzer built a number of submarine diesels; the largest, the 6M375, was installed in three vessels that entered service between 1924 and 1926. The largest submarine to enter service with Sulzer diesel engines, however, was the French *Surcouf*, in 1934; its two engines with 7,600-hp combined output gave a surface speed of 18 knots. The most powerful Q-type engines of this period were two 8Q65 engines,

First Sulzer 6U23 submarine engine *(top)*, and 4,000-hp 9Q51N.

The experimental US Navy submarine, *S-2*, with twin Busch-Sulzer engines.

each of 6,500 hp, in the French submarine depot ship *Jules Verne* in 1933.

The next phase of submarine engine development began around 1936 with the application of airless fuel injection. The 1,500-hp 3Q51ES crosshead-type research engine of 1931 was the first engine to run in Winterthur with an all-welded structure. Another engine, the DQ68, was intended as a prototype for a 1939 battleship, which, if built, would have had quadruple-screw machinery totalling 264,000 hp, employing eight 14-cylinder in-line engines in pairs in tandem direct drive on each shaft.

A number of 6,900 hp 10QDC51 engines were built before World War II, for the *Roland Morillot* class of large submarines. But, owing to the war, none of them was completed. Development work also continued with single-acting two-stroke engines for naval applications. Turbocharging and welded construction were expected to give greater powers at lighter weight. This concept led in 1956 to the V-form VQA42 type that was developed at the request of CCM and the French licensees. The design, however, was superseded by the gas turbine for future naval applications.

French submarine *Casabianca*, with Q-type engines *(top)*, and the Polish minelayer *Gryf*, the only naval ship with Sulzer SD-type engines.

Airless fuel injection

A major drawback of early marine diesel engines was the blast-injection system with its need for large, high-pressure air compressors, which demanded much maintenance. With blast injection, the fuel oil was blown into the cylinder by a blast of compressed air at a pressure of 65 to 75 atmospheres. In 1892, Rudolf Diesel had envisaged the direct, mechanical injection of liquid fuel. Practical difficulties in constructing a precise pump and injecting the fuel rapidly, however, led to the expedient of blast injection, which gave good atomisation of the fuel. This remained in use into the 1930s, but the air compressors gave much trouble, required additional maintenance effort and took around seven per cent of the engine power.

The development of airless injection is a good example of higher technology being introduced into engine design. The high peak pressures generated in the airless injection system were feared because of the need for a stronger camshaft drive, with high stresses for cams and tappet rollers. Suitable materials and manufacturing techniques had to be found to achieve the required accuracy and surface quality in cams, rollers, pump plungers, pump sleeves, and injector needles. Means of generating the extremely high injection pressures, and the techniques for distributing and atomising the fuel and for exact control of the quantity and timing of each injection, had first to be perfected.

The difficulties with the early trials were not surprising because in those days nothing was known about the relationships between fuel injection pressure, length, and diameter of high-pressure pipe, pressure waves, number and diameter of atomiser holes, jet length, the ability of a fuel jet to penetrate throughout the compressed air charge in the cylinder, atomisation, duration of injection, and combustion chamber geometry.

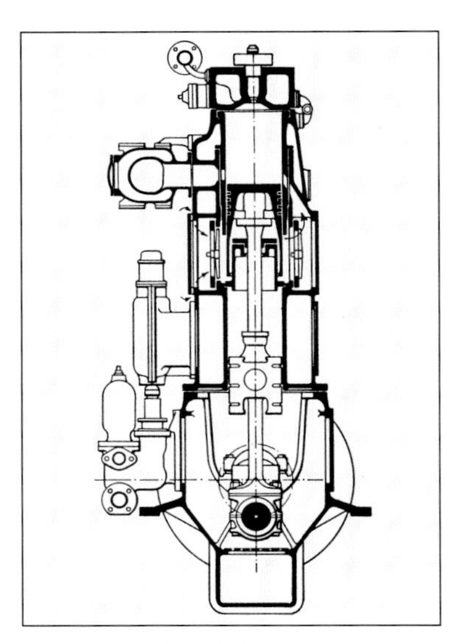

RK engine (1925).

It is not surprising that airless injection was first employed by Sulzer in relatively small RV engines, of very simple design, with no inlet or exhaust valves, and no valve gear. Air was drawn into the crankcase through automatic disc valves in the crankcase doors and discharged to the engine cylinder through a single row of scavenge air ports in the liner. The exhaust ports were opposite the scavenge ports.

From the earliest test engines, Sulzer developed the RV and later the RK series. More than 1,900 RK engines were built from 1925 to 1955, with a combined output of over 332,500 hp. At first, the RK engines had indirect fuel injection with pre-combustion chambers and, after 1932, direct fuel injection was employed. Other improvements were made on the basis of the service experience, and the extensive

A ZG9 in an air-raid shelter in Geneva.

A 3ZG9 opposed-piston test engine installed in a five-tonne lorry.

testing of research engines in the Ludwigshafen, Winterthur and CCM works. For example, the specific outputs were increased by some 40 per cent in the RKP type by fitting additional scavenge air pumps to operate in parallel with the crosshead pumping.

Overall, the RK types were designed for maximum simplicity with regard to construction and attendance, the greatest possible reliability, economy in operation, and low manufacturing cost in series production. As with the RV types, they were widely employed for marine propulsion, in direct-reversing form or with reversing gear, as marine auxiliaries and for stationary drives in factories and electricity power plants.

Airless injection also made possible another series of small, high-speed diesel engines developed by Sulzer, the opposed-piston ZG type from 1935 to 1949. Their configuration was quite a break from past designs, with horizontal cylinders and opposed pistons driving through rocker arms to a single crankshaft located beneath the cylinders. They were notable for very good power/weight ratios. More than 1,000 ZG engines were produced for stationary installations or marine applications, with a relatively few of them manufactured for traction duties in lorries, tractors, and railcars.

Large-bore airless injection

Although direct injection had been first tested at Sulzer in 1915, it was systematically studied from 1924 onwards. The first results on a test engine definitely demonstrated the superiority of direct injection over pre-combustion chamber systems. The full-load fuel consumption dropped more than six per cent over the use of a pre-combustion chamber system. Investigations with direct injection in large-bore engines, begun in 1925, however, were not so convincing that the well-tried blast injection system could be immediately abandoned for large-bore Sulzer engines.

To circumvent the difficulties with jerk-type pumps much effort was put into accumulating systems, in which fuel was first stored under pressure before delivery for injection. A number of elegant accumulating systems were developed by Sulzer for airless injection in the 1920s. In the end, however, accumulating systems were found to be too complicated.

The next generation of jerk pumps had a fixed length of stroke and the quantity of fuel delivered was regulated by varying the start of injection, with a constant end of injection. The suction valve was held open until the appropriate moment for the commencement of injection, according to the load setting.

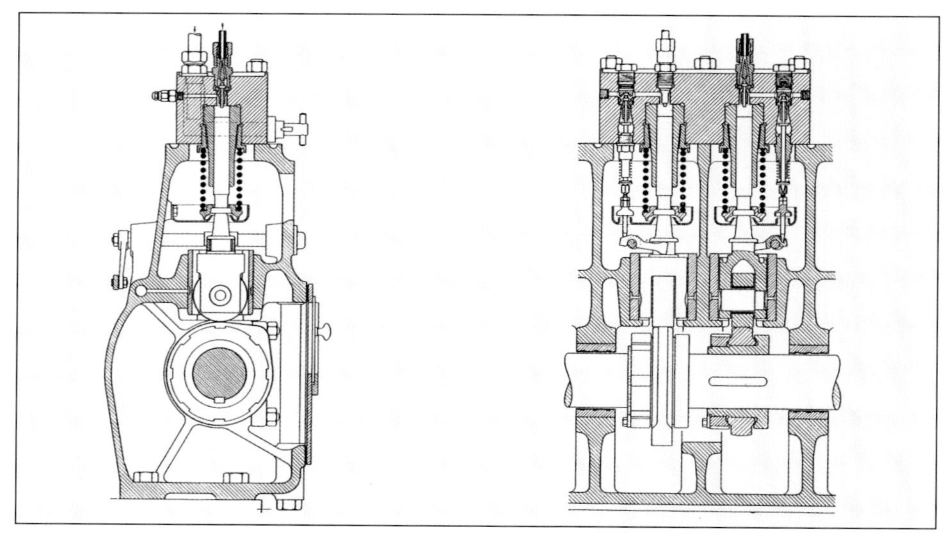

Fuel pumps for the SD60 and SD72 after 1941, with variable injection start.

Single control, at either the beginning or the end of injection delivery, was used in the early stages of development when turbocharging had not yet been applied and thus when compression pressure was practically constant at all loads. Double-valve control of fuel pumps was introduced in the RSD58 of 1950 to give flexibility in injection control to suit the variability in ignition qualities of heavy-fuel bunkers. With the addition of turbocharging, the disadvantages of late injection at part load became more pronounced owing to less air delivery at low load.

The SD-type engine

The SD airless-injection single-acting two-stroke marine diesel engines of 1931/32 were manufactured until 1964 by Sulzer Brothers and the licensees. More than 752 SD engines were built with an aggregate output in excess of 3,650,885 horsepower. It is notable that, although only a modest number of engines were built in the 1930s, the SD type engines strongly benefited from the sudden jump in demand for newbuildings after World

War II. The first SD type engines to enter service were the twin-geared 7SD56 units from Werkspoor used in 1934 to re-engine the three12,700-dwt Dutch cargo ships *Manoeran*, *Madoera* and *Mapia*.

The first diesel-driven cross-Channel packet, the 3,120-grt Belgian *Prince Baudouin*, was powered by twin 12-cylinder SDT58 engines built under license by John Cockerill with a combined output of 17,000 horsepower directly driving twin screws at 268 rpm. These were the first 12-cylinder Sulzer two-stroke engines. At the time, she was the world's fastest motorship, having reached 25.23 knots during sea trials in August 1934. She was employed on the day service between Ostend and Dover, taking six hours for the double crossing to meet train connections at both ends of the route. The *Prince Baudouin* led to a long line of Belgian cross-Channel packets, which served until February 1997.

Isipingo **(*top, left*) had early SD engines. Twin 12SDT58, 8,500-hp engines (*top, right*) were installed in the Belgian packet *Prince Baudouin*.**

Lower fuel consumption
A landmark in diesel technology was reached in June 1934 when the remarkably low fuel consumption of 149.2 g/hph at full load was achieved by an 8SD72 engine during its official shop test in the Winterthur works. This was the first of a pair of engines for the 13,650-dwt reefer *Dorset* of the Federal Steam Navigation delivered in November 1934. The *Dorset* and her sister *Durham* were both built by Workman Clark in Belfast, who also built the engines for the *Durham*.

The low fuel consumption figures were confirmed by Professor G. Eichelberg of ETH Zürich in tests on the second engine. The low consumption was attributed to a combination of factors: high mechanical efficiency; only about five per cent of engine power was needed to drive the scavenge air pumps; good fuel atomisation and

The 13,650-dwt refrigerated vessel, *Dorset*, powered by two 8SD72, Winterthur-built, engines.

distribution in the combustion space; and the high maximum cylinder pressure of 60 atmospheres.

The test results from the *Dorset* engines inspired Sulzer Brothers to build a single-cylinder SD72 research engine to obtain a sound basis for further development. It was started in July 1936, and was still employed for research in the early 1950s. Overall, the research on the 1SD72 engine was comprehensive and clearly made a great contribution to advancing the technology of two-stroke diesel engines in Winterthur. The wealth of results were rapidly applied in improvements to the SD engine types. They provided some simplification, with lower manufacturing costs, reduction in maintenance time, lower piston withdrawal height, shorter engine length, and less weight.

The French-built *La Marseillaise* (1949) was powered by three 11SD72 engines totalling 25,000 horsepower.

Burning heavy fuel oil

A key development for large marine diesel engines around 1950 was their adaptation to burn the cheaper, heavier fuel oils that became more prevalent as marine bunkers after World War II. Over the years, these fuels were known under various names, including boiler fuel, boiler oil, residual fuel, Bunker "C," and high-viscosity fuel, but today the usual term is simply heavy fuel oil.

The use of heavy fuel oils has become a crucial element in the remarkable operating economy of large diesel engines. It made them fully competitive with steam plants, not only with lower fuel consumption but also their ability to burn the same fuels. Fuels with poor combustion characteristics had been used extensively in numerous Sulzer stationary engines from the earliest years, and included coal tar oil as well as heavy fuel oils. Much greater wear and fouling of pistons and cylinder liners diminished the economics of heavy fuels. The first Sulzer engine to run on heavy fuel, in 1912, was a 120-hp, four-stroke trunk-piston unit in Argentina.

After the Second World War, serious efforts were made in the marine industry to solve the problems of burning heavy fuel oils in large diesel engines, as were being burned under boilers. Pioneering work was undertaken by John Lamb of the Anglo-Saxon Petroleum Co, which included demonstrations on board the 12,100-dwt tanker *Auricula* from 1946 onwards. The key to satisfactory operation on such fuels came in the mid-1950s with the availability of cylinder lubricants containing enough alkaline additives to neutralise the acidic combustion products and thereby to bring wear rates down to those usual with diesel oil. The passenger liner *Willem Ruys* was utilised in 1953 for the first sea trials of Shell's new alkaline cylinder lubricant.

In the early work with heavy fuels, however, it was recognised that the best performance could be given by engines designed specifically for operation on poor-quality fuels. At the same time, the fuel injection system was modified in various ways and provision was made to allow hot fuel to be circulated to keep the fuel pumps and fuel injection valves warm, while fuel pipes were traced with steam or hot water. Engines could thus be readily started and stopped on heavy fuel, providing the fuel heating and circulation were properly maintained.

Research into the behaviour of engines running on heavy fuels was undertaken in Winterthur between 1950 and 1952 on the 1TS72 research engine. In parallel, experience was being gained at sea with SD engines converted to run on heavy fuels.

In the late 1970s, bleak forecasts of considerable deterioration in the quality of heavy fuel oils led Sulzer Brothers to undertake a long-term research programme to investigate all aspects of fuel quality, which included a series of in-service trials from 1979 onwards on board the LASH ship *Bilderdijk*, powered by a Sulzer 9RND90 engine. Using sophisticated instrumentation and a versatile fuel system, it was possible to evaluate thermodynamic conditions and wear in two of the cylinders by running them on selected heavy fuel with various properties.

The experimental 1TS72, a two-stroke trunk piston engine *(right)* in Oberwinterthur.

Alternate fuels

Apart from the commonplace middle distillate fuels termed gas oil and marine diesel oil, the fuels used in the past century have included various petroleum products from kerosene to residual fuel oils, as well as other liquid fuels, including coal tar oil and peanut oil, and gases such as producer gas and natural gas (methane).

Rudolf Diesel envisaged that his engine would burn coal, but he used liquid fuels from the beginning. He tested his engine on coal in 1899, only after his main research programme was completed, but without success. Many other researchers have since tried to run diesel engines on coal and particular attention was given in the USA to the problem after the 1973/74 oil crisis. In 1978, the US Department of Energy contracted Sulzer Brothers and Thermo Electron Corporation to investigate the use of coal-derived fuels in low-speed two-stroke diesel engines. A further research programme was initiated in 1982, and later tests were made with synthetic liquid fuels derived from coal and slurries of microscopically-ground coal in diesel oil or water.

The shortage of liquid fuel oils during World War II renewed interest in using natural gas and producer gas from coal and wood. In 1941, a Sulzer 3DD22 was converted successfully to dual-fuel operation. Several other DD-type four-stroke engines were converted for running on gas as dual-fuel engines, or with spark-ignition. In the early 1950s a 6TF48 trunk-piston two-stroke engine was also adapted as a dual-fuel engine running on low-pressure natural gas with about five per cent fuel oil for pilot ignition, but only a few such engines were put into service.

***Venator*, the first LNG carrier to burn cargo boil-off gas in a diesel engine.**

Researchers also saw that dual-fuel versions of the large two-stroke engines could be applied to the propulsion of liquefied natural gas (LNG) carriers burning boil-off gas from the cargo. Tests began in 1964 on the 1RSA76 research engine. In 1969, a 6RD76 engine was demonstrated in Winterthur running with five per cent pilot fuel oil. However, the only low-speed two-stroke marine engine in the world to enter service, in 1972, running on natural gas was the Sulzer 7RNMD90 engine in the Norwegian 29,000 m^3 LNG carrier *Venator*. The limited quantity of boil-off gas available called for at least 30 per cent fuel oil, with which the engine developed 14,000 hp. The full output of 20,300 hp at 122 rpm could be developed on 48 per cent gas, 52 per cent oil.

High-pressure gas injection developed in 1986 was found to be more promising. By injecting high-pressure gas (250 bar) in the same manner as liquid fuel after compression of the air alone, the engine output is not restricted by knock and the full "diesel" output can also be obtained on gas with pilot-oil injection.

In 1972, two Spanish-built 115,000-dwt Conoco tankers were equipped with AESA-Sulzer 23,200-hp 8RND90 main engines, auxiliaries and boilers that burn crude petroleum. The ships ran between single-pont moorings, thus saved on the logistical cost of taking on normal bunkers.

The RS-type engine

The RS types, the first Sulzer two-stroke engines specifically designed for running on heavy fuel oil, were introduced to allow greater outputs on single shafts. Of the features in the RS to facilitate heavy-fuel operation, the most important was the piston rod stuffing box, or gland, which separated the crankcase, running gear, and system oil from the cylinder space and its harmful combustion residues. Although the lantern at the foot of the cylinder of the SD type with its set of scraper rings served a similar purpose, it was not as effective as the gland, which sealed well on the smaller diameter of the rod.

The first two 10RSG58 engines were ordered in 1950 for a twin-geared 9,000-hp single-screw installation in the 10,134-dwt refrigerated cargo liner *Middlesex*, delivered in 1953. The *Middlesex* was one of four refrigerated ships built between 1953 and 1955 for the New Zealand Shipping Co and Federal Steam Navigation that were powered by twin-geared Sulzer RSG58 engines. The *Northumberland* also had twin 10-cylinder engines, and the *Otaki* and *Essex* each had twin 5,400-hp 12-cylinder engines. The first RSD76 engines were ordered in 1952, and installed in the 7,200-dwt cargo ships *Duquesne* and *Frontenac*. The first RSD76 type from Winterthur was a 10-cylinder unit of

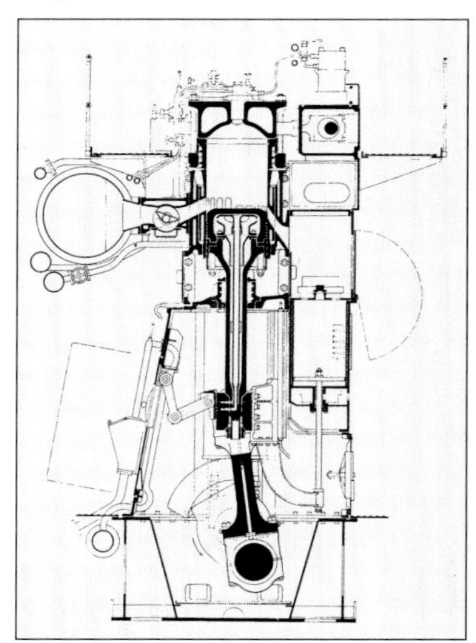

RS58 engine (1950).

The first RS-type engine built in Winterthur was this 10RSG58 for the *Middlesex*.

The second Sulzer 10RSG58 engine for *Middlesex* was built by Alexander Stephen.

9,000 hp built in 1954. It was installed in the 10,750-grt Royal Rotterdam Lloyd cargo ship *Wonosobo*, delivered to the line in 1955.

Twin 12RSG58s with electromagnetic slip couplings and reduction gearbox.

Supercharging

Exhaust-gas turbocharging has been the greatest single technical contribution to diesel engine progress. It removed the constraints of natural aspiration, enabling great increases in power output, reductions in engine size and weight, and considerably helped the diesel engine to displace the steam turbine and become the compact, economical prime mover it is today. So important has turbocharging now become that the development of large diesel engines is largely dependent upon further advances in turbochargers. But there was some 20 years between its adoption in four-stroke diesel engines in the mid-1930s and its application in large marine two-stroke engines in the mid-1950s.

The concept of supercharging, or pressure charging to increase engine power, was considered from outset. Rudolf Diesel even tested supercharging in his second engine of 1896, using the pumping effect of the piston underside to precompress the charge air. He returned, however, to supercharging in 1906 with his "precompression combustion process," in which the working cylinder was charged with highly-compressed air from a reciprocating air pump.

In a joint development with Diesel, it was tested by Sulzer Brothers on one 1D40 research engine modified to two-stroke operation, with scavenge air supplied by the other 1D40 research engine rebuilt as a two-stage air pump and driven by an electric motor. These tests were continued under the supervision of Alfred J. Büchi (1879–1959), after he joined the firm in 1909. They showed that supercharged operation was practicable with mean indicated pressures up to 19.6 bar, and formed the basis for the first diesel locomotive in 1912.

The double-row scavenging arrangement of Sulzer two-stroke diesel engines in 1909 was an early implementation of supercharging. The upper row of scavenge air ports remained open after the exhaust ports had been covered by the piston, thereby enabling the cylinder to be charged up to the full pressure of the scavenge air. The slight overpressure of the scavenge air thus gave a modest supercharging effect.

This principle was utilised in the aftercharging of Sulzer ZA and ZDA stationary two-stroke engines to boost engine output by some 20 per cent for peak load generating duties. Tests were first made in 1924 on a 4Z60 stationary engine. The additional boost air was supplied from an extra pump mounted on top of the usual crankshaft-driven scavenge pump, and separately ducted to the cylinders. Mechanically controlled valves then admitted this additional air through the upper row of scavenge ports after the piston had covered the exhaust ports. The aftercharging system was brought into operation automatically when the engine load exceeded 75 to 80 per cent. The only marine engines to use this form of supercharging were the four 10ST68 engines of the French liners *Aramis* and *Félix Roussel*, when they were converted in 1936 to airless fuel injection. The engine outputs were raised from 5,800 to 7,350 horsepower each.

In the early 1930s, Sulzer Brothers also applied the aftercharging concept to the DD-type four-stroke engines. It was a compound admission or topping-up system. The cylinder heads had the usual inlet and exhaust valves but, in addition, there was a row of boost-air ports around the bottom of the cylinder liner. At the end of the suction stroke, the inlet valve was closed before the ports were uncovered by the piston. The incoming boost air then topped the cylinder air charge up to the boost pressure. The boost air was delivered by a reciprocating compressor driven from the engine. Engine output was increased by up to 30 per cent, while maximum cylinder pressure was practically unaltered. The concept was dropped, however, once exhaust-gas turbochargers of suitable size, and with adequate efficiency, became available.

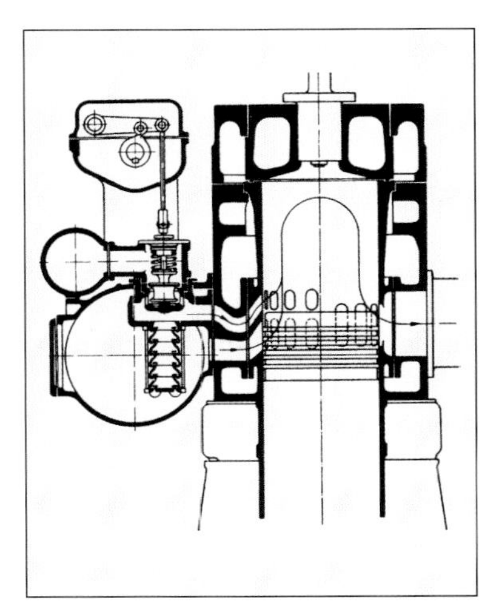

Sulzer aftercharging system.

Turbocharging

Exhaust-gas turbocharging had a long gestation period. Alfred J. Büchi, the most famous name associated with turbocharging, patented in 1905 a turbocompound engine in which a turbocompressor, a highly-supercharged reciprocating engine, and an exhaust-gas turbine were all connected together. Other configurations were included in his 1906 US patent. His 1909 paper, however, also showed the turbine driving

The first completely turbocharged Sulzer two-stroke test engine, a 6TA48.

The first turbocharged RSAD76 (1956) was installed in the *Marne Lloyd*.

only the compressor as a free-running machine with shaft output from the reciprocating engine alone, as in the usual arrangement of a turbocharged engine. Professor Auguste Camille Edmond Rateau (1863–1930) in France also took out a patent for the turbocharging process in 1916, and applied exhaust-gas turbochargers to aircraft engines during World War I to maintain their power at high altitudes.

The concept of a highly-supercharged four-stroke engine running at a moderately high speed, however, was not welcomed by Sulzer Brothers. The firm preferred the two-stroke diesel engine with which it was making good business in ship and power plant applications. After a long dispute, Büchi left the firm in 1918 but, after working at Howaldtswerke in Kiel, he returned to Winterthur in 1920 as manager of Sulzer's patent department.

Once the predicted potential of turbocharging had been confirmed, the Büchi system was eagerly adopted by many engine manufacturers for four-stroke engines. In March 1936, Sulzer Brothers concluded a license agreement with the Büchi Syndicate for the general application of exhaust gas turbocharging according to the Büchi patents.

Once the turbocharged four-stroke diesel engine was accepted by the market in the mid-1930s, considerable attention was given by the principal engine designers to solving the much greater problems of two-stroke turbocharging. Not being self-aspirating, two-stroke engines require greater air flows than equivalent four-stroke engines to ensure a large quantity of trapped air for complete combustion and to limit thermal loads.

At some stage in the two-stroke cycle, it is usual to have both inlet and exhaust ports and valves open so that the incoming air can thoroughly scavenge the exhaust gases from the cylinder. The resulting greater air flow, however, reduces the exhaust temperatures and the exhaust energy available to drive the turbocharger turbine. At the same time, scavenging efficiency is very sensitive to exhaust back pressure. The pressure after the cylinder must be sufficiently below the scavenge air pressure for the desired air quantity to flow through the cylinder. The overall result is that turbochargers need to have a much higher efficiency for two-stroke operation. Even when the problem was resolved for high loads, part-load running remained difficult without additional air blowers or pumps.

The necessary improvements in turbocharger design were eventually aided through parallel developments in gas turbines in the 1940s, while the principal engine manufacturers were undertaking lengthy development programmes, supported by extensive test-bed trials, to study the implementation of turbocharging.

In Sulzer's case, experimental work on two-stroke turbocharging began with supercharging tests on single-cylinder research engines, the 1QD42 in 1934 and the 1SD72 in 1939, using separately-driven scavenge-air pumps while simulating the turbine by throttling the exhaust gases. Researchers gained further experience with engines running at increased mean effective pressures with the 10STA68 engine conversions in 1936 for the *Félix Roussel* and *Aramis*.

The first complete turbocharged engine tested at Sulzer beginning in 1941 was a 6TA48 cross-scavenged two-stroke trunk-piston engine. Its unsupercharged output of 2,100 hp at 250 rpm was raised by 25 per cent with pulse-pressure turbocharging. In 1946, it became the first turbocharged Sulzer two-stroke diesel engine to enter regular service, when it was installed to drive a generator in the powerhouse of the Winterthur works.

Further turbocharging tests with cross-scavenged two-stroke engines, both trunk-piston and crosshead types, were continued in the 1950s. A 1RS58 research engine was even modified to uniflow scavenging to settle the question of whether cross-flow or uniflow scavenging was better. Four exhaust valves were arranged in the cylinder head around a central fuel injection valve. As expected, uniflow scavenging showed a small gain in specific fuel consumption but, at the time, it was not thought adequate to merit the exhaust valves, valve gear, and more complicated cylinder head casting of the uniflow-scavenged engine.

Turbocharging was applied to SD and RSD engines in around 1955. Constant-pressure turbocharging was applied to SD60 and SD72 two-stroke engines in the SAD versions. The first turbocharged SAD72 to enter service was a Uraga-built 5,400-hp six-cylinder unit in the cargo ship *Dowa Maru* in December 1956. A 6SAD60 of 4,100 hp was ordered from the Winterthur works in January 1956 and installed in the *Hoegh Ailette*, commissioned in 1958.

The first turbocharged RSAD76 engine was a seven-cylinder unit of 8,400 hp built at Mitsubishi's Kobe works in May 1955. It entered service in December 1956 in the cargo ship *Montevideo Maru*. Early in 1956, a 9RSAD76 also began running on the test bed in Winterthur. Developing 11,700 horsepower, it was installed in the 11,520-dwt Dutch cargo liner *Marne Lloyd* delivered in April 1957. (The first turbochargers designed and manufactured by Sulzer were smaller units ordered in 1939 with four-stroke traction engines. Licensees building Sulzer diesel engines were free to use either Sulzer or BBC turbochargers. Sulzer turbochargers were also manufactured by licensees in Japan. However, in 1969, Sulzer discontinued the design and manufacture of its own exhaust-gas turbochargers.)

The 11,520-dwt *Marne Lloyd* powered by an 11,700-hp Sulzer RSAD engine.

High-pressure turbos

In parallel with the development of low-pressure turbocharging for large two-stroke diesel engines, Sulzer Brothers also carried out an extensive research programme into the supercharging of two-stroke engines at high scavenge air pressures. The research involved a number of uniflow-scavenged, opposed-piston, two-stroke engines of the ZG and G types. The opposed-piston engines were developed under Dr Kurt Retschy (1895–1960), who was appointed head of research in the diesel department in 1935. Since 1924, he had worked on Junkers opposed-piston engines. He originated some important concepts that proved to be very forward looking, namely the rotating piston, streamlined plate-type automatic non-return valves for scavenging systems, bore cooling, and backward-curved turbocharger compressor blades. Rotating pistons, for example, which reduce piston ring wear, were employed in opposed piston engines of 90- to 320-mm bore in the 1930s and 1940s, and subsequently in Sulzer four-stroke engines up to 650 mm bore. They have been standard on all Z-type engines since the mid-1960s.

The opposed-piston 6G18 (1939).

In highly-supercharged engines, as the boost (or scavenge-air) pressure is increased, the proportion of work done by the exhaust turbine rises, together with the power absorbed by the air compressor. At a scavenge pressure of five to six bar, the three work components are equal. The reciprocating engine and the compressor can then form a power gas producer, and all useful output can be delivered by the exhaust-gas turbine. This is the basis of the free-piston plant.

An adaptation of turbocompounding was commercially adopted in the 1980s for the RTA-series two-stroke engines in the form of the Sulzer Efficiency-Booster System (EBS). Turbochargers then had sufficient surplus capability that about 10 per cent of the exhaust gas could be diverted to a turbine coupled through an integral "power take-in" gear to the engine crankshaft. Operating in parallel with the exhaust-gas turbochargers, the power turbine was supplied with gas only when the engine was running above about half load. Instead of increasing engine output, it was employed to reduce the engine's fuel consumption.

By 1996, power turbines were no longer being offered. There was then less interest among shipowners in the added complication and extra cost. In any case, turbocharger performance was being more fully utilised within the engines themselves as mean effective pressures rose to 19 bar, and maximum cylinder pressures to 150 bar.

Exhaust-powered turbine and power take-in gear on a 6RTA58.

Rail traction engines

The most nostalgic development in Sulzer four-stroke engines was, in the eyes of many, the rail traction engines built between 1912 and 1977. Sulzer's diesel traction business was something of a cradle for the diesel department, where many engineers began their professional lives before serving in the wider marine and power plant world. The main personality associated with the early years of Sulzer diesel traction was Adolf Brunner. He joined Sulzer Brothers as an apprentice. He collaborated in design of the "Thermo-Lokomotive" engine and the 6LV26 railcar engines. In 1924 he was appointed manager of the diesel locomotive group, from which came numerous locomotives and traction engines.

Although Sulzer Brothers never manufactured locomotives in Winterthur, the firm often acted as main contractor for locomotives designed by itself and, in the early years, also fitted diesel engines and electrical equipment into the mechanical portions manufactured by subcontractors.

The Sulzer diesel traction business began with a valveless two-stroke engine, the 1,000-hp V-four in the world's first diesel-engined rail locomotive, the 1912 "Thermo-Lokomotive" built for the Prussian railways.

The "Thermo-Lokomotive" (1912) with direct-drive V-four engine and compressed-air bottles *(left)*.

The main 4LV38 direct-reversing two-stroke engine directly drove two axles through coupling rods on disc cranks on the ends of the crankshaft. To start the train, the engine had to run on air (from bottles charged by a compressor) from standstill up to some 10 km/h, or usually more, to reach a minimum running speed for the engine. Only then could the starting air be shut off, and fuel applied to the engine so it became self sustaining.

The "Thermo-Lokomotive" can best be described, alas, as a fiasco. It was run only for trials and these were halted by World War I. The high pressures generated as the cylinders started firing, while still being fed with highly compressed air, imposed severe stresses on engine components, with resulting fracture of the crankshaft and cracking of cylinder covers.

The correct path proved to be diesel-electric transmission with quicker-running four-stroke engines. In 1914, Sulzer supplied five diesel-electric rail cars to the Prussian and Saxony railways. Each was propelled by a 6LV26 V-form four-stroke engine of 200 hp at 440 rpm, which set the pattern for all further developments in the traction field.

The diesel-electric locomotive (1937) for hauling French express trains.

Among the highlights that followed were the direct, airless injection LTD engines of all-welded construction. The 6LF19 version was employed in two prototype "Red Arrows" rail cars built for the Swiss railways in 1936 by SLM. Another major landmark in Sulzer traction engines was the use of twin-bank engines in three 4,400-hp diesel-electric locomotives, one for France in 1937, and two for Romania in 1938. The vehicles were of different designs, but similar in being double-unit locomotives with each unit housing a Sulzer 12LDA31 turbocharged engine with a rating of 2,200 hp at 700 rpm.

The French locomotive was used in express PLM service between Paris and the Riviera, whereas the power of the two Romania locomotives was set to hauling trains over a 1,054-metre pass in the Transylvanian Alps. As the then most powerful diesel locomotives in the world, these were another milestone, and the high point of Adolf Brunner's career.

In 1956 Romania ordered six locomotives with 2100-hp 12LDA28 engines. At the same time Romania bought the rights to build the locomotives and engines without paying further fees. This had the astounding result that by 1987 more than 3,500 locomotives and their engines had been manufactured in Romania. These are widely employed in Poland, China, North Korea, and Bulgaria, as well as Romania itself.

More important for the Sulzer traction business, however, was the modernisation of British Railways (BR) with new diesel locomotives from 1957 to 1968. Of the 3,090 mainline locomotives then purchased by BR, 1,397 were powered by Sulzer diesel engines. They were LDA28 engines of six-, eight-, and 12 cylinders and most were built in a joint venture with Sulzer UK by Armstrong-Vickers in Barrow-in-Furness.

The most powerful locomotive with a Sulzer engine was the British *Kestrel* with a 16-cylinder LVA24 engine of 4,000 hp. After being used for trials, it was sold to the USSR in 1971.

The 4,000-hp, high-speed *Kestrel*, which originally served in the UK.

Other four-stroke engines

From the D-type engines of 1903, Sulzer produced a continuous succession of four-stroke diesel engine designs for driving generators and water pumps, or industrial drives. After Sulzer Brothers took over Schweizerische Lokomotiv-

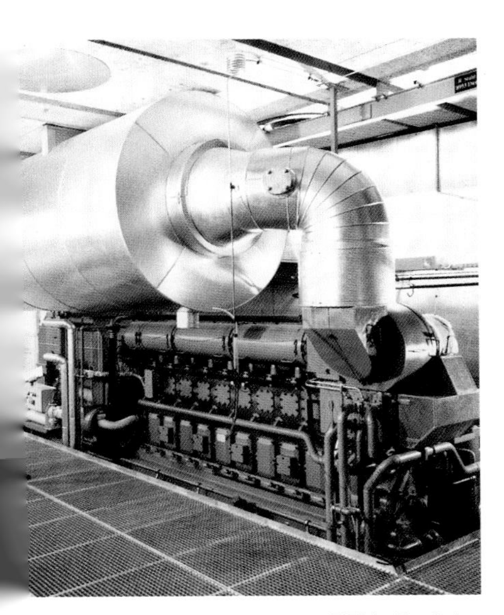

A Sulzer 9S20 genset at ETH, Zurich.

und Maschinenfabrik (SLM) in 1961, it was decided to develop a new four-stroke engine, the Sulzer A25, which was introduced in 1966.

For the initial rating of the A25 the engines had two-valve cylinder heads, whereas four-valve heads were introduced for the higher rating. In 1973, the A25 was supplemented by the AS25 of 245 hp/cylinder and the A20 of 125 hp/cylinder. A heavy-fuel version of the AS25 was added in 1976, featuring two-part pistons with forged steel crowns, bore-cooled cylinder heads, and other appropriate modifications. It was uprated to 270 hp/cylinder to burn marine diesel oil. They were joined in 1982 by the AT25 giving 300 hp/cylinder.

After signing a licensing agreement in 1980, Waukesha Engine Division of Dresser Industries Inc in the USA adapted AT25 engines to spark-ignition, lean-burn versions to run on natural gas. Later they have developed a larger version. They are also manufacturing the standard liquid-fuelled AT25. The various A-type engines have been used for diverse applications, including marine auxiliary, marine propulsion, stationary generating sets in base-load and standby plants, and rail traction. By the end of 1997, some 6,165 engines had been built or were on order, with an aggregate output of 6.47 million horsepower.

A new design, the S20, with the same cylinder dimensions as the A20, was introduced in 1988 to provide a competitive range of in-line engines for ship auxiliary duties. The output of S20 engines have since been increased, and a four-cylinder model added.

The excursion vessel *Lausanne*, with twin Cegielski-built 6S20s, was once chartered for the Sulzer licensee meeting in 1996.

Two-stroke trunk-piston
One of the more successful series of Sulzer diesel engines were the T types, of which more than 2,075 engines, amounting to 2.96 million horsepower, were manufactured between 1932 and 1983. The T type originated as a successor to the small RK crosshead two-stroke engines which, although serving well, involved high production costs and were rather tall.

Built in Winterthur and at many licensees, T-type engines were employed in all types of vessel in deep-sea, short-sea, coastal and river operation, including tugs, fishing craft, and whale catchers. In most single- or twin-screw propulsion installations the engines were direct reversing. They were also built in large numbers with designation TF for power plants ashore.

T-type engines were installed in many interesting vessels, including a series of passenger ships and ferries built from the mid-1930s by William Denny & Brothers Ltd on the Clyde. The Canadian icebreaking ferry *Abegweit*, which had a 12,000-hp diesel-electric plant with 16 Dominion-Sulzer 6TE29 engines in eight tandem generating sets, was quite unusual. For geared drives and diesel-electric installations, Sulzer developed another series of two-stroke trunk-piston engines, the M type, in the 1940s. As it was for merchant ships, the M type used cast-iron construction rather than the cast steel of its predecessor Q types.

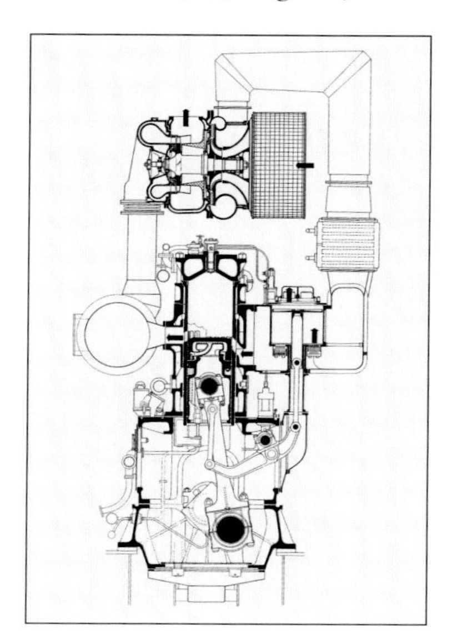

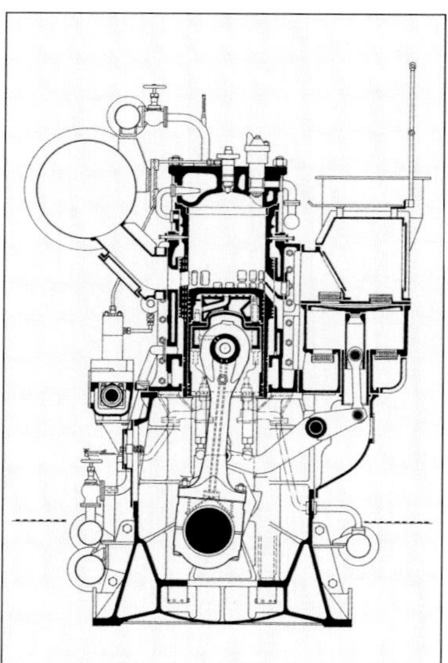

T-type *(top)* and M-type.

A 10MPD51 in Winterthur before installation on the ferry *Prinses Josephine-Charlotte*.

The Belgian car ferry *Prinses Josephine-Charlotte* was delivered in 1949 with twin Winterthur-built 10MPD51 engines totalling 7,600 hp. She was followed by four more Belgian cross-Channel ferries each equipped with twin Winterthur-built 12MD51 engines, of 9,600 hp combined output. The M51 type was also extensively used by Wärtsilä AB in 14 icebreakers and two icebreaking cable-layers built in Finland for the USSR, all with diesel-electric plants.

The most powerful of the icebreakers were the five *Moskva*-class polar vessels built between 1960 and 1968, each with eight Wärtsilä-Sulzer 9MH51 engines totaling 27,000 hp. In the Netherlands, Royal Schelde had built 24 of the 8MH42 for the series of six *Lena*-class icebreaking cargo ships delivered between 1954 and 1957.

From two- to four-stroke
In the mid 1950s, there arose questions about whether the two-stroke or the four-stroke cycle was more appropriate for future engine designs. Accordingly, in 1956, Sulzer built two 4UV32 V-form trunk-piston research engines, that could be adapted to two-stroke or four-stroke operation, with either 400- or 500-mm stroke. The engines were used in a wide range of investigations until 1968, including tests of pulse and constant-pressure turbocharging. One of the test engines reached 24.5 bar mean effective pressure, running as a two-stroke using two-stage turbocharging.

The research on the 4UV32 engines laid the groundwork for the Z-type trunk-piston engines in the early 1960s. Of the first, the ZV30, only three were built for navy projects. The second, the Z40, found its first customer in Wärtsilä, as a successor to the Sulzer M51 two-stroke engines they were building in Finland for large Soviet icebreakers. (The two-stroke engine had excellent torque characteristics for handling the power-surge demand of running through ice.)

The first engines to enter service, in 1967, were four nine-cylinder units, of 16,400 hp

An experimental 4UV32, adaptable to either two- or four-stroke operation.

combined output, in the 8,168-grt Baltic ferry *Finlandia*. She is still in service today, as the cruise ship *Costa Playa*. Wärtsilä subsequently supplied Z40 two-stroke engines for seven large Soviet icebreakers delivered between 1974 and 1981. The most powerful were the three *Yermak* class, each with nine 12-cylinder engines, totalling 41,400 horsepower.

For the main markets, a four-stroke version of the Z40, designed to burn heavy fuel oil, appeared in 1970. V-form models were added to the range and the output was increased step-by-step. The ZA40 type was announced in 1981; it was joined in 1985 by the longer-stroke version, the ZA40S. In 1971, an agreement was made between Sulzer and MAN to design much larger four-stroke engines. It resulted in the 12V65/65 prototype being started in the Winterthur works in 1975. The market for this engine type was not then very encouraging and, in 1979, the agreement was ended. In 1995, the ZA power range was extended by the 1,630 hp per cylinder ZA50S, which has been based on the successful experience with the ZA40S.

In little more than a quarter of a century, Sulzer has sold almost 1,700 Z-type four-stroke

Carnival Cruise Lines' *Fantasy* is powered under tropical skies by four 12-cylinder and two 8-cylinder ZA40S engines.

engines. The Sulzer Z-type engines have found good application as prime movers in specialised vessels, many with multiple propellers, such as small passenger vessels, ferries, Ro/Ros, dredgers, tugs, fishing boats, and icebreakers, running in a direct-drive mode through a reduction gearbox, or through a diesel-electric drive.

The diesel-electric system allows the greatest flexibility in locating the engines, therefore one extremely popular application is aboard cruise ships, where the smaller the engine room the greater the hull volume that can be given over to accommodations.

The astonishing list of cruise ships that use Sulzer four-stroke drives includes the six Carnival Cruise Line

P&O Cruise Lines' *Sun Princess* is powered at sea by four 16-cylinder ZA40S engines.

***Canmar Kigoriak*, a powerful Canadian icebreaker, does its solitary duty with two 12ZV40/48 engines.**

Fantasy series; Disney Corporation's *Disney Magic*; five Holland America Line cruise ships including *Statendam* and *Rotterdam*; the four Princesses of P&O Cruises (one of which, *Grand Princess*, has six 16-cylinder ZA40S engines); and NYK Line's *Crystal Symphony*. The engines have also found exciting application in the highly successful *Superfast* ferries running between Greece and Italy, and in the new US Coast Guard icebreaker, *Michael A. Healey*.

Both the ZA40S and the ZA50S engines have the unique Sulzer rotating piston, driven by a ratchet mechanism inside the piston. The rotating piston, patented in 1936, solved the problem of wear caused by the cyclical side forces exerted on the trunk piston's head by the pivoting piston rod. By rotating the piston steadily through each cycle, the wear is distributed evenly, allowing much longer hours of operation between overhauls.

Although the Sulzer brand four-stroke engine may never share in the romantic limelight of its larger two-stroke brethren, it is an important secondary player on the stage of diesel engine history, and the Sulzer four-stroke brand is certainly far from moribund.

The RD series

The power range of Sulzer low-speed two-stroke diesel engines was considerably increased in 1956 with the addition of the RD76. It was the first Sulzer two-stroke engine designed from the outset for turbocharging, and conceived as a homogeneous series of geometrically similar engines of five bore sizes. In the RD, the scavenge air receiver, turbochargers, and scavenge air coolers were combined into one group at the side of the engine, simplifying the structure and giving shorter connections for the scavenge air and exhaust gases. Scavenging was changed from cross-flow to loop, thereby improving scavenging efficiency.

Whereas turbocharging on the RSAD76 boosted its output by some 30 per cent, the RD76 ultimately gave about 70 per cent more power than an unsupercharged engine. It was followed in 1957 by the RD90 of 2,000 to 2,300 hp/cylinder, giving a maximum of 27,600 hp from a 12RD90 model. Other design changes from the RSAD type to the RD types included the use of water cooling for the pistons, instead of oil, to cope with the somewhat higher thermal load.

From the late 1950s onwards, the sizes of ships, particularly tankers, increased continually, which gave the impetus for ever more powerful engines and especially the super-bore of 900 mm or more. The first 12-cylinder RD76 engines, supplied by licensees Iino Shipbuilding Co and Royal Schelde, went to sea in 1960 in the 47,252-dwt tanker *Kakubo Maru* and the 32,000-dwt tanker *Barendrecht*. The first 12RD90 followed in 1964 from IHI in the 96,500-dwt tanker *Kirishima Maru*.

The speeds of cargo liners was also rising. For example, the two graceful 11,380-dwt Union-Castle cargo liners *Southampton Castle* and *Good Hope Castle* delivered in 1965 were powered by twin Wallsend-built 8RD90 engines, each power plant delivering a total output of 35,200 hp and giving the ships a service speed of 22.5 knots. The fast cargo liners of the mid-1960s, however, were the last of their breed. They were superseded by container-ships, which came to require even more powerful engines.

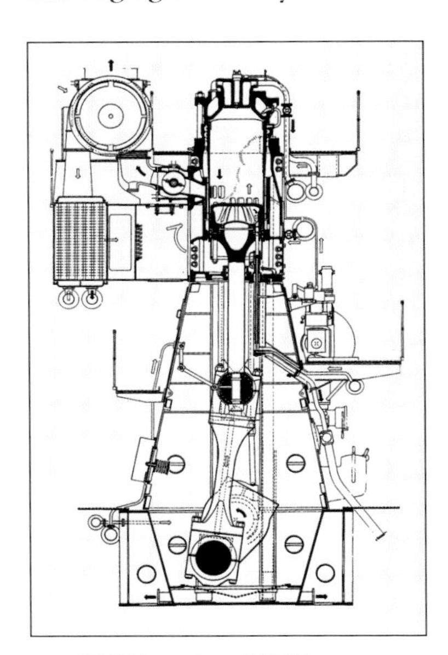

RD90 engine (1957).

Designed for turbocharging, a 9,600-hp 6RD76 is tested in Winterthur.

Stationary power plants
Turbocharged RD two-stroke engines were soon employed for electricity generation in many base-load power stations. The first installation was in a 6.7-MWe 8RF68 generator set commissioned at Clifton Pier power station of the Bahamas Electricity Corporation in 1963. Its highly satisfactory performance led to three two-stroke engines being installed in 1983 and a fourth in 1989.

An impressive array of eleven Sulzer two-stroke engines in Bong Range mine, Liberia.

Sulzer RF68-type low-speed engines were also installed at the Bong Range iron ore mines in Liberia, Mexico, and Mecca. Larger 9RF76 sets of 12-MWe output were installed in Sahara and the Canary Islands. The introduction of the Sulzer RNF stationary engine design, directly adapted from the RND marine engine, led to extensions at Bong Range and Mecca, while two 9.25-MWe 8RNF68 sets were installed between 1978 and 1980 at King Tom station in Sierra Leone.

By 1977, Bong Range had become the world's largest diesel power station with a total capacity of 95 MWe from four 9RF68 and seven 9RNF68 engines. By the end of 1984, the 11 engines had accumulated some 961,259 running hours, with the longest-running engine exceeding 124,585 hours. The Bong Range, however, was surpassed by the 156-MWe Vazzio power station of Electricité de France on Corsica. The station has eight generating sets, each with an 8RNF90M of 27,200 hp.

The RTA-series engines have also been extensively applied to electricity generation, with 32 engines of 824,710 hp in seven countries. Two 9RTA58-engined sets were installed on Guernsey in the Channel Islands in 1987 and 1993, joining three 9RNF68M engines from 1979 to 1982. Nearly half the total, however, have been installed on various Greek islands: 15 engines totalling some 149-MWe output.

An RNF90M generating electrical power.

The RND series

The RND series was introduced in 1967, the year that the Suez Canal was closed giving the impetus to the development of VLCCs suitable for oil transport on the longer route round the Cape of Good Hope. The first type in the new series was the 1,050-mm bore RND105, an unprecedented cylinder size, combined with the higher mean effective pressure to give a dramatic increase in unit output to 4,000 hp per cylinder, or 48,000 hp from a 12-cylinder unit. Improved turbocharger efficiency made possible a change to constant-pressure turbocharging for the RND engine series; this allowed higher pressure with lower fuel consumption, without excessive thermal loadings. Constant-pressure turbocharging also simplified engine design, eliminating the exhaust rotary valves of the RD engines and giving more freedom in the numbers of cylinders.

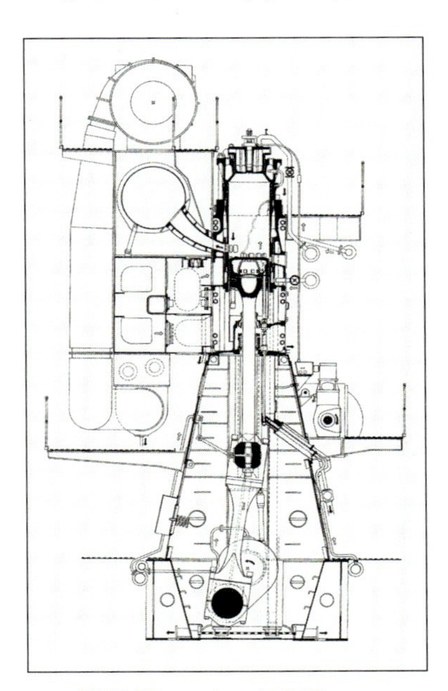

RND90 engine (1967).

Demand for higher powers tended to precede the turbocharger performance available. In the mid 1960s, Sulzer undertook extensive test-bed trials of two-stage turbocharging, using an 8RD68 engine, which also included comparisons of pulse and constant-pressure turbocharging. Sulzer performed more two-stage turbocharging trials in 1975 on an 8RND90 engine, in which adequate specific air flows were maintained at mean effective pressures up to 16 to 17 bar. More importantly, the two-stage turbocharging trials demonstrated the capability of the RND90 to develop higher power, paving the way for the RND90M of 3,350 hp/cylinder.

Though two-stage turbocharging has been shown to be impressive, and a few such engines have even gone to sea, in general it has presented added complication and cost. Besides, two-stage turbocharging has so far been rendered unnecessary through continued turbocharger development, and modern high-efficiency turbochargers for two-stroke crosshead engines already run at some 19 bar with only single-stage supercharging.

One of the world's most powerful motorships in 1971, *Thames Maru*.

During the late 1960s and early 1970s the rapid rise in the size of containerships increased demand for more powerful propulsion. Although some owners preferred steam turbine plants, such as the twin-screw 81,120-hp machinery in OCL's five 2,300-TEU *Liverpool Bay* class, there were soon notable large diesel installations. In any case, the 1970s oil crisis killed the marine steam turbine for all merchant ships except large liquefied natural gas carriers.

Although the RND105 was employed in a number of single-screw installations, there was soon a need for twin-screw plant to reach the required propulsion power. The 1,800-TEU *Kurobe Maru*, *New York Maru*, *Kiso Maru*, and *Svendborg Maersk*, delivered between 1972 and 1974, each had twin 12RND90 engines of 69,600 hp combined output.

The swing from steam turbines to diesels after the crisis resulted in some 80 large containerships being delivered or refit with Sulzer RND90M engines. The most powerful of these newbuildings was the 2,388-TEU *Thames Maru* of Mitsui OSK Lines in 1977. She had twin 12RND90M engines totalling 84,000 hp for a service speed of 26.5 knots. She remained for some years one of the world's most powerful motorships. However, high fuel costs led to her engines being permanently de-rated to 54,200 hp and her cargo capacity being increased to 2,900 TEU.

The RND-M series
The RND-M series, introduced in 1976, was an improved version of the RND, with a 15 per cent increase in power output (the range was 6,940 to 40,200 hp). In addition to extending bore cooling to the single-piece forged steel cylinder heads as well as the cylinder liners, the RND-M design incorporated a number of improvements developed from RND experience, including a new combustion chamber, modified fuel injection system, accumulator-type cylinder lubricator system and a re-designed piston for better running behaviour. Standard ISO metric threads were adopted throughout. To give a better margin against bearing fatigue strength, the RND-M crosshead was fitted with thin-walled aluminium-tin bearing shells and provided with separate high-pressure lubrication.

Cylinder blocks for an Allis-Chalmers 12RND90M engine destined for an APL 2,480-TEU containership (1982).

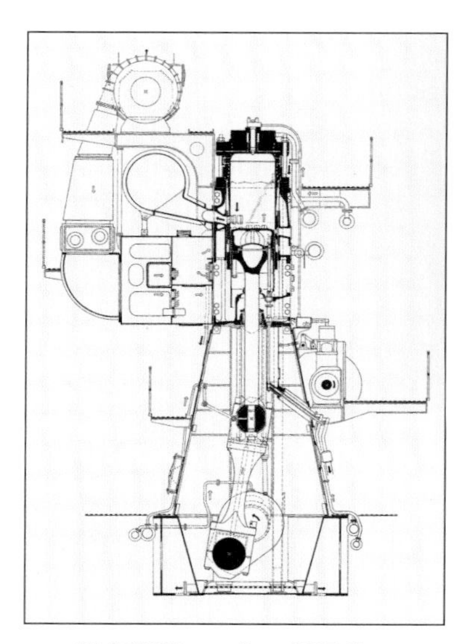

RND90M engine (1976).

After the quest for ever higher unit outputs during the preceding decades, development in the 1970s was characterised by simpler, yet more sophisticated solutions. A good example of this is the step-by-step introduction of bore cooling, devised at Sulzer Brothers in the mid-1930s. But, production application began only in 1962 with the cylinder liners of Z-type engines, and continued to RND liners, RND-M cylinder covers, RL pistons, so that all combustion chamber components of the RL, and now in the RTA series, are bore cooled. By segregating mechanical and thermal design criteria, bore cooling is the best way of controlling stresses and strains in these critical components, despite the major increase in power output and maximum cylinder pressures.

Three 12-cylinder RND90M engines were manufactured in the USA for the 2,480-TEU C-9 containerships of American President Lines (APL). Built at Avondale in 1982 and 1983, the *President Lincoln*, *President Washington* and *President Monroe* are the largest containerships ever built in the USA and they were the first to be equipped with diesel engines. The 43,600-hp engines were subcontracted by Sulzer Brothers to Allis-Chalmers Corp in Milwaukee, Wisconsin. Subsequently, in 1985, APL purchased second-hand two other containerships with 12RND90M engines. They also took delivery of five 4,340-TEU newbuildings, the C10s, in 1988. Each vessel is powered by a Sulzer 12RTA84 engine of 57,000 horsepower.

APL's *President Lincoln* enters San Francisco Bay driven by a Sulzer 12RND90M engine.

The long-stroke RL series
Following the 1973/74 oil crisis, the marine industry made comprehensive efforts to reduce fuel consumption. For example, the trend to lower propeller revolutions for higher propulsive efficiency was met in 1977 by the introduction of the longer-stroke Sulzer RL engines in which stroke/bore ratios were increased from about 1.67 in the RND-M series to 2.1.

At the same time, fresh optimisation of combustion and scavenging processes, combined with the use of higher-efficiency turbochargers, gave further reductions in specific fuel consumption.

To reach higher thermal efficiencies, development was also aimed toward increasing the maximum cylinder pressure without compromising engine wear and reliability. Modest increases were adopted in succeeding low-speed two-stroke engines, rising from 76 to 80 bar in the RD, to 85 bar in the RND, and reaching 94 bar in the RND-M. (Most recently, the Sulzer RTA48T-B and RTA58T-B engines have attained maximum operating pressures of 150 bar, or double that of the RD some 40 years before.)

Studies of slow-steaming engines had shown the fuel-saving possibilities in maintaining the maximum cylinder pressure at its full-load value over a wider load range, within the limitations imposed by bearing load capabilities.

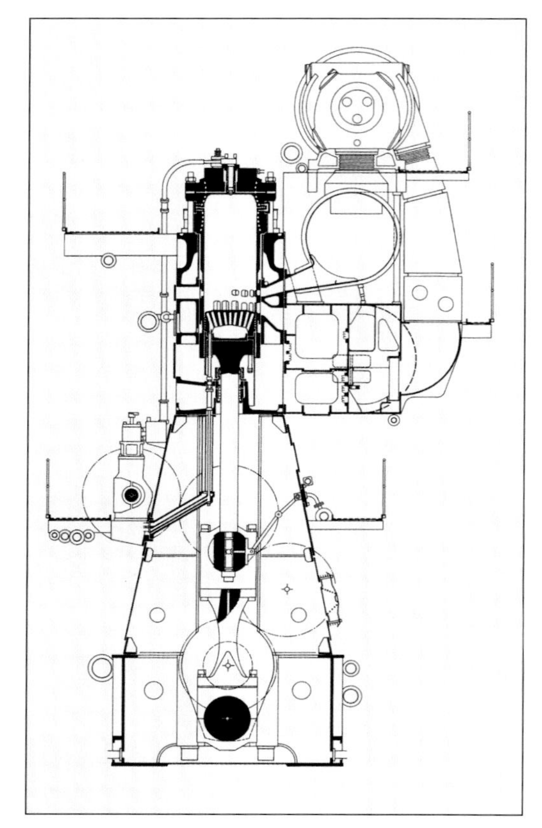
RL90 engine (1977).

Sulzer RL-series engines were thus the first engines to be equipped with a variable-injection timing (VIT) mechanism as standard. Continued now in the RTA series, this gives load-dependent superimposed control of the timing of the whole fuel injection phase and thereby keeps the combustion pressure virtually constant at its maximum value down to 85 per cent load and then automatically reduces it as a function of engine load. It also incorporates manual adjustment to compensate for delayed combustion with poorer-quality fuels. Sulzer's valve-controlled fuel injection pump readily suits modification for VIT.

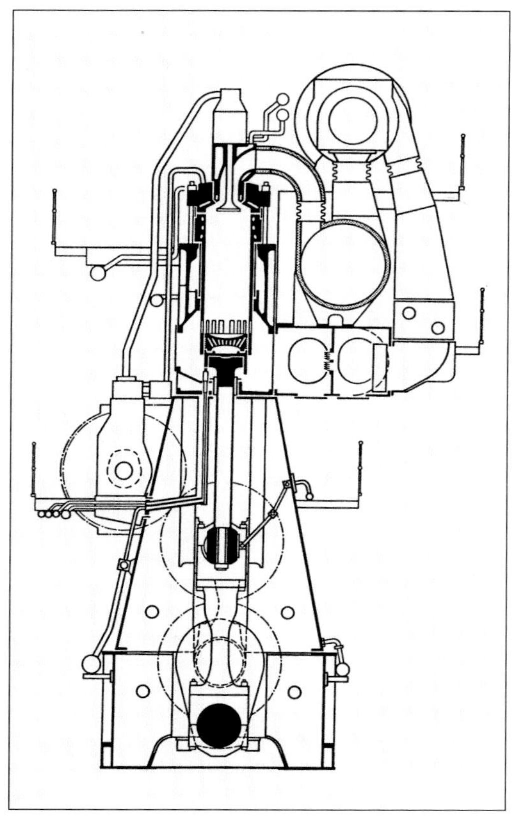
RTA84 engine (1982).

Uniflow scavenging
Further improvements in overall propulsive efficiency were made possible with the Sulzer RTA low-speed engine series announced in January 1982. Compared with the RL types, the new designs offered both lower propeller speeds and much reduced specific fuel consumption. The RTA design achieved remarkable market success. Within two years, it had attracted orders for 394 engines with a total output of more than four million horsepower.

A radical change in design concept was made in the RTA. It was the first large Sulzer two-stroke diesel to use uniflow scavenging. The change of direction had clear technical origins. Analysis by Sulzer of ship types, ship sizes, service speeds, fuel costs, engine prices, engine dimensions, and propeller design, indicated the need for lower shaft speeds. Lengthening the stroke/bore ratio is the only solution for low-speed engines, permitting a simple, cost-effective direct propeller drive. But increasing the stroke/bore ratio has an adverse influence on engine cost and dimensions. It was thus considered reasonable to increase the stroke/bore ratio from 2.1 in the RL series to 2.9 in the RTA. At this substantially increased ratio, the traditional valveless port-scavenging system is thermodynamically at a disadvantage, and the change to uniflow scavenging with single poppet exhaust valves was therefore quite logical.

The RTA series was introduced in six bore sizes, the broadest range of crosshead engines ever produced by Sulzer. Their nominal outputs spanned from 2,000 to 54,000 hp at speeds between 196 and 65 rpm, with specific fuel oil consumptions as low as only 117 g/hph at 85 to 90 per cent load in derated versions, equivalent to a overall efficiency of 53 per cent.

Complete layout flexibility in the RTA series facilitates the choice of optimum engine ratings for virtually any ship. Contract ratings for any RTA model can be freely selected within a layout field extending over a certain range of power and speed. By retaining the nominal maximum cylinder pressure at whatever contract rating is selected, this layout field affords reductions in fuel consumption of up to several grams from the MCR level.

The first RTA engine to enter service, in November 1983, was a 9,600-hp 6RTA58, built by Mitsubishi and installed in the 38,000-dwt bulk carrier *Ocean Great*.

The full-scale uniflow-scavenged R2 test engine that served as the prototype for the Sulzer RTA series.

Today

We have seen, to this point, the technical history of the Sulzer engine — how inventive minds and skilled hands in the lovely tree-lined town of Winterthur changed Rudolf Diesel's rough invention into an extended family of technologically remarkable (if not all successful) engines. Perhaps it is time here for a bit of political history, to get temporary relief from the mean effective pressures, revolutions and horsepowers of the past, and look at the Sulzer diesel brand today.

Throughout its years, as is apparent from the preceding pages, the Sulzer brand diesel engine has been actively manufactured in Winterthur and by more than 100 licensees around the world. It is not unusual, in this highly specialised world, for an industry to depend heavily on licensing. After all, the history of entrepreneurship is filled with clever inventors who did their mental work in a laboratory, but then built no more than an unsophisticated prototype of their invention. They had to find a sympathetic, equally clever organisation, with solid machinery (and no-less solid financing) to mass-produce the invention and bring it to market, for which they were paid a royalty. For this incentive alone, many men have worked long and hard to build that elusive better mousetrap. Think of Thomas Edison, surely the most prolific inventor of all time. He was awarded more than 1,000 patents in his lifetime, yet he hardly manufactured a light bulb, gramophone, microphone, or kinetoscope.

But the diesel engine business is unique: It was virtually *founded* on the concept of licensing, by Rudolf Diesel himself. As you have seen, in the late 19th century Diesel sensed correctly that it would be prohibitive for him to invest in tooling up a new factory to produce an engine that was at best experimental, at worst possibly unbuildable. His only hope of realising his dream of replacing the steam engine with a more efficient one based on the principle of compression ignition, was to license his know-how and patents to companies that already had machinery and production capability in place. Thus, between 1893 and 1898, the Sulzer Brothers fell in among a group of nearly a dozen companies (in Switzerland, Germany, France, Sweden, Scotland, Denmark, Belgium, and the USA) who signed license agreements with Rudolf Diesel to produce his new engines.

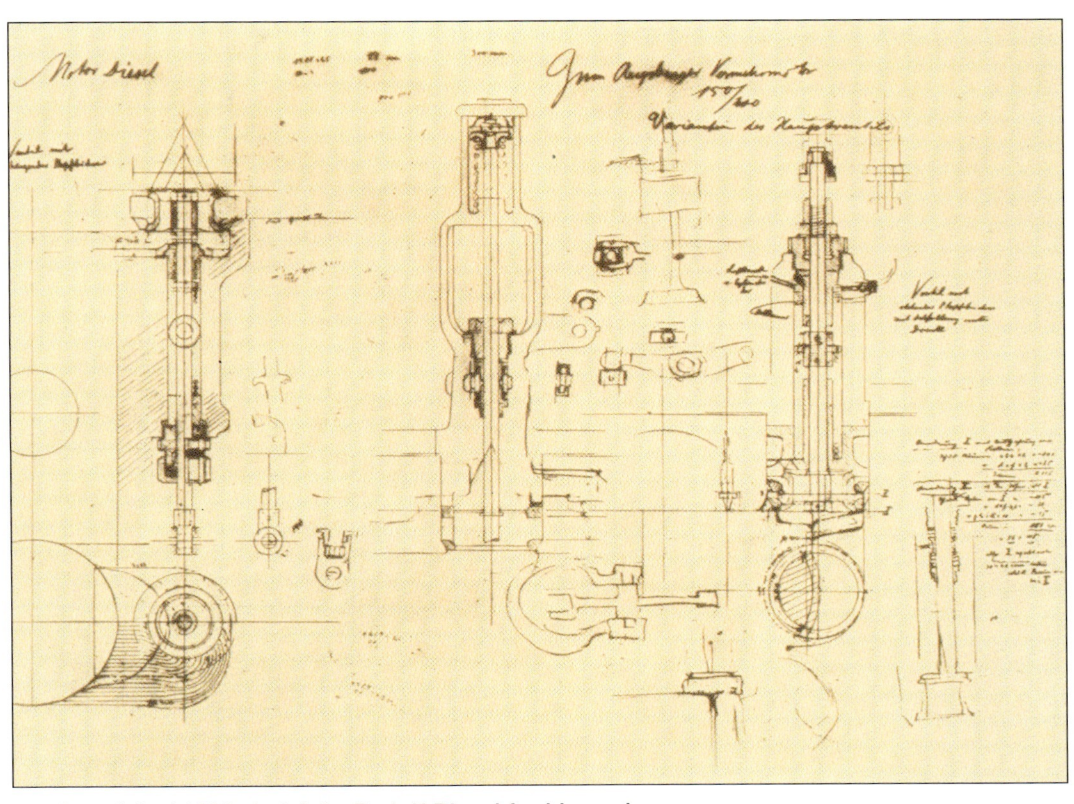

An original 1892 sketch by Rudolf Diesel for his engine.

Sulzer Brothers was a natural choice, not only because of its earlier connection to the young Diesel when he was a student, but because of its well-established manufacture of steam engines, compressors, and other proven industrial machinery. For example, in the 1830s, Sulzer Brothers made equipment for the textile mills set up along the meandering rivers near Winterthur. When the mills found that the hydropower they drew from the rivers no longer provided sufficient energy for their expanding manufacturing, Sulzer provided them with steam engines. Then, Sulzer found it logical to put steam engines on lake and river boats, until Rudolf Diesel offered the world something better. (By 1947, when Sulzer ceased manufacturing steam engines in Winterthur the company had delivered more than 6,500 engines with a

The single cylinder of a Sulzer 1D20 engine No. 21, built in 1903.

combined output of some 1.6 million horsepower.)

Some years after Rudolf Diesel sold his patents and rights to the Allgemeine, Sulzer Brothers signed an agreement with the Allgemeine, which granted the firm permission to sell its brand not only in Switzerland, but virtually anywhere in the world. Sulzer, though continuing to build engines in Winterthur, transformed itself into a licensee *and* a licensor. From that shining moment, nearly 100 years ago, the structure of the diesel engine business changed forever.

Sulzer's ability to license its brand was a brilliant success, largely because it was able to manufacture its engines close to world markets, without significant capital investment — factors which are still fundamental to licensing in most industries today. Over the next 80 or 90 years Sulzer signed more than 90 license agreements for manufacture of its engines in more than 120 diesel works in more than 30 countries on four continents, while it continued to manufacture diesel engines in Winterthur.

But by the 1980s the firm found itself in exceptional circumstances: It was doing research, development, and design for engines it believed right for the world market; but its Winterthur production had fallen to less than five per cent of the demand for those engines.

And with the exponential growth in demand for marine diesels, it became abundantly clear to Sulzer's management that, for many reasons, Winterthur was no longer a convenient place to build engines for large ships, particularly the new breed of tankers, bulkers, and containerships that have come to rule the seas.

The most obvious among the reasons for Winterthur's "obsolescence" in large two-stroke engine production was geography: By its isolation in central Europe — 450 km over the Alps to Trieste, 500 to Genoa, and 650 to Rotterdam — it generally is not a good place to manufacture goods for heavy industry; it has no raw materials, no local markets.

And Winterthur is a quintessentially landlocked town. Sulzer had to import its steel, cast and machine engine parts, assemble the engines, test them in town, dismantle them, and move them in sections to the ports by railway and river, for shipment around the world. (Inasmuch as local Swiss railway tunnels cannot accommodate loads wider than four

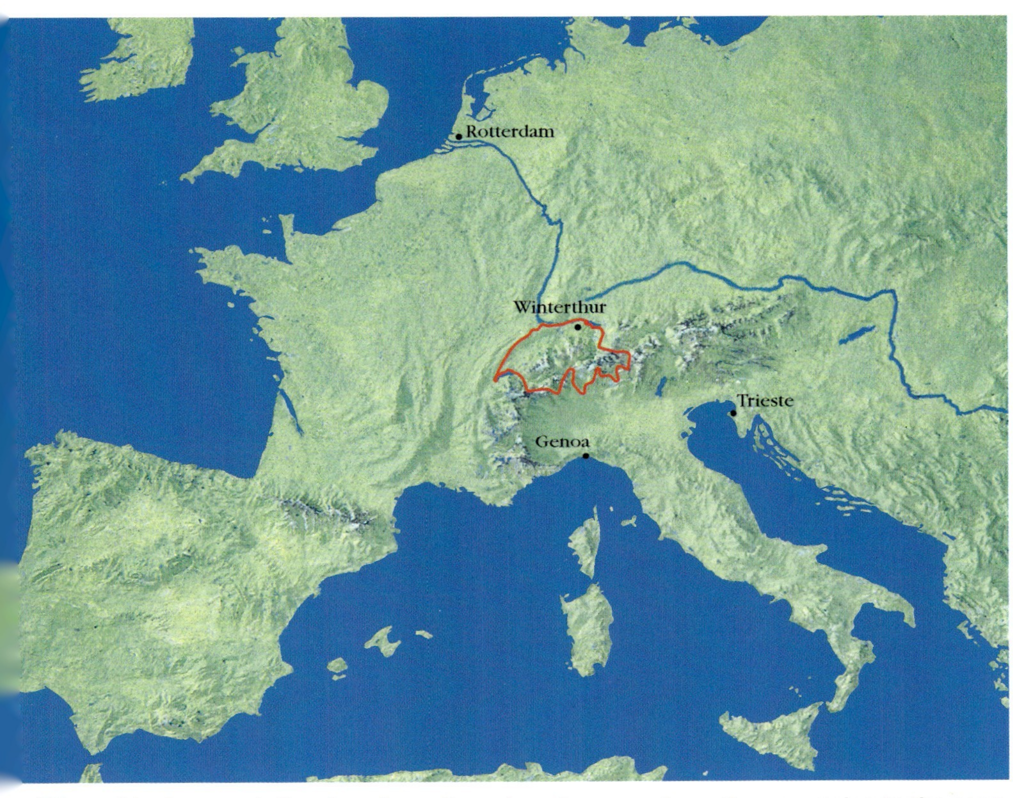

When shipping great diesel engines, it can be a long way from the mountains to the seas.

metres, shipment of large engines by rail for loading onto barges on the Rhine at Basel was problematic at best.) Also, by its remoteness, the factory could certainly not respond as quickly as it should to new orders.

As a result of these bottlenecks, Winterthur engines were costing at least 10 per cent more than anyone else's. Yes, there were loyal clients who demanded only a "Winterthur engine," and who were surely prepared to pay for it (there still are some today), but they were a small minority in a marketplace becoming ever more dominated by cost pressure and price consciousness.

So, Winterthur found itself too far from the prime users of its products, world-class shipping companies — and farther still from the shipyards that built their ships, most of which were already in Asia. Indeed, 80 per cent of world shipping is currently built in Japan, Korea, and China; the concept of regularly shipping 1,000-tonne engines from the heartland of Switzerland to Asia (and widening all those railway tunnels!) borders on the absurd.

And, perhaps the final blow for Sulzer was that, by the peculiarity of the licensing life, the company experienced growing competition, even potential conflict of interest, between its own engine production and that of its many licensees. (For example, Winterthur had a foundry, so Sulzer traditionally designed engines heavily dependent upon cast parts. But, after World War II many licensees, particularly those large operations in East Asia that are tied to shipyards and share their construction philosophy, came to prefer welding or forging of large parts. Sulzer necessarily changed to welded construction to accommodate them where possible.)

By the mid-1980s, Sulzer's management saw the handwriting on the wall; it was clearly time to stop production in Winterthur entirely. Despite some pressure from the local government to keep the manufacturing going, to protect hundreds of jobs, it was impossible for Sulzer to go on. The company elected to suspend manufacturing in Winterthur, to depend in the future wholly on licensees, and to apply its not insubstantial license income to more intensive research, development, and design of future engines.

The logic of Sulzer's decision to exist by licensing alone, even if made a bit late, was unassailable. It kept the brand available. And it created a spirit of cooperative risk sharing with licensees: Winterthur bears the development risk; the licensees bear the manufacturing and sale risk. Finally, licensing is a very elegant tool for a creative company to transfer technology and get paid for bright ideas, without having to invest in building and operating great factories all over the world.

Sulzer Brothers shipped its last Winterthur-built marine engine to the Howaldtswerke-Deutsche Werft, in Kiel, Germany, in the late summer of 1986. It was a 12,580-hp 6RTA62 for the 31,205-dwt containership *Norasia Al-Mansoorah*. In the end of December 1988, Sulzer shipped the very last engine manufactured in Winterthur. It was a 15,200-hp 8RNF68M destined for the Clifton Pier power station of the Bahamas Electricity Corporation.

***Norasia Al-Mansoorah*, the beneficiary of the last marine diesel engine shipped from the Winterthur works.**

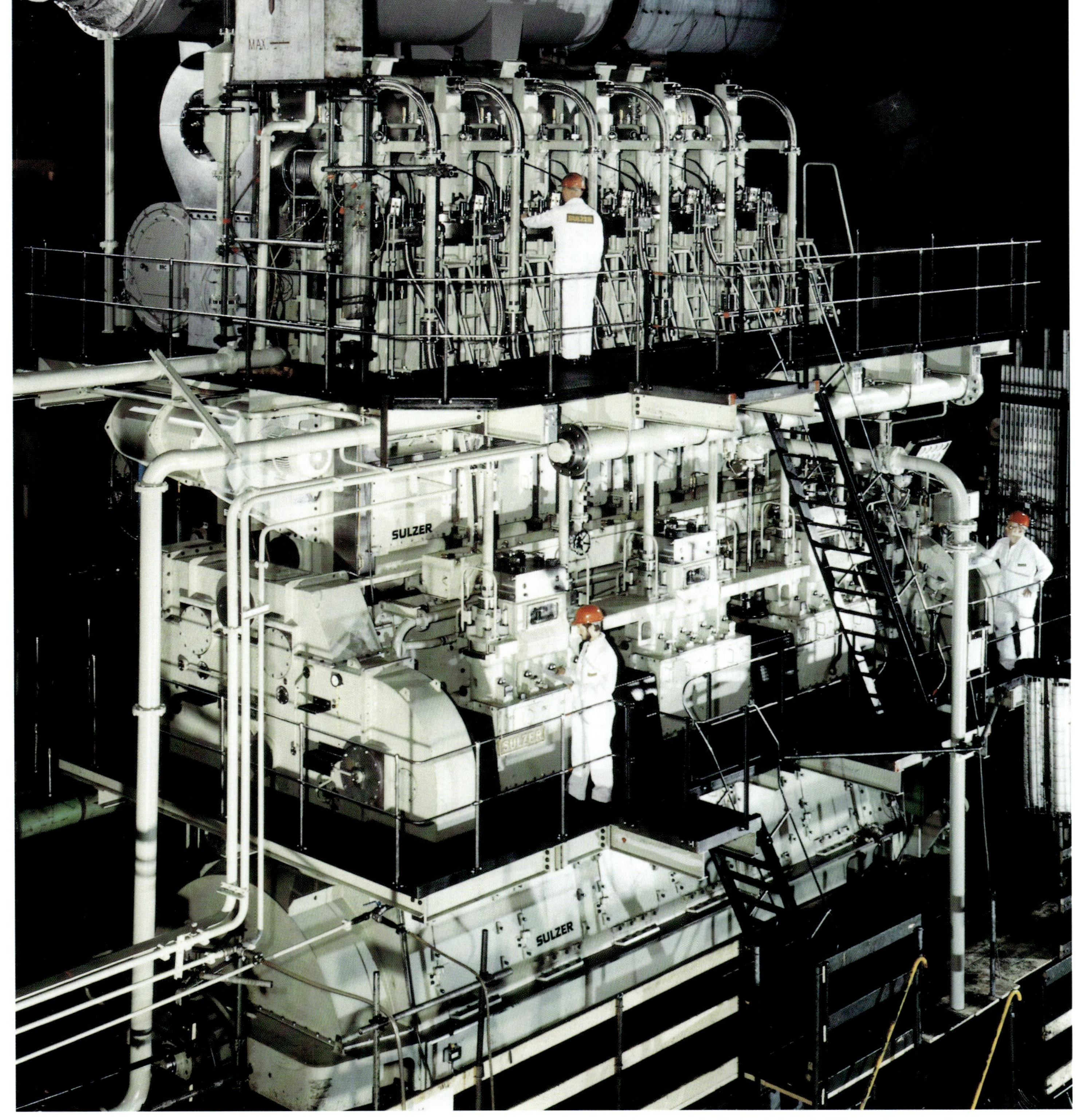

A six-cylinder RTA62 has the sad distinction of being the last marine engine ever built in Sulzer's Winterthur works.

The making of Sulzer's last Winterthur engines was to the outside world not much more than a news item in the marine and engineering presses. But to many people within the bosom of the company — people who had dedicated years of energy, insight, and sensitivity, even their entire adult lives, to the development of the Sulzer diesel — it was a sad day, as many of them felt that the company had "licensed" their jobs away. One might ironically call this event "the day the music died," because it ended nearly a century of glorious diesel engine manufacturing in the gracious, tree-lined city of Winterthur.

The situation was new for many Sulzer people, and the rhythm of the company necessarily changed. They could no longer gaze out their office windows to see a mighty engine being set on a railway wagon. More deeply, the role played in their lives by licensees now took on a greater weight and importance. As Albert Senhauser, a retired vice president for license contracts, expressed it: "When we stopped making engines in Winterthur, and became dependent on licensees, the most difficult thing we had to do was learn how to listen."

The modern Sulzer office tower stands as a noble backdrop to the original engineering works.

There was, however, an even more difficult challenge lying ahead for the women and men of Sulzer. This time it was the challenge of survival. From 1988, when it ceased manufacturing, the Sulzer brand was built exclusively by licensees around the world. But on 1 July 1989 the Sulzer Brothers management made a stunning turnabout — nine years short of the 100th anniversary of building their first engine for Rudolf Diesel, they set up their diesel engine business as an independent company, MBS Dieselmotoren AG.

This was the opening of a new era and the beginning of a new set of challenges for this hundred-year-old brand. People with a clear view of the diesel engine marketplace felt that the sale of the Sulzer diesel department was a healthy one, a certain means for the brand to thrive as an independent entity. But, again, old-timers, many of whom grew up in Winterthur, sat at Sulzer drawing boards and toiled at Sulzer test beds, felt that Sulzer had sold its own heart, its own reputation, which was absolutely centred on the diesel engine. (It was time for loyal employees to take comfort in the great Swiss playwright, Friedrich Dürrenmatt, who said that when the world turns bad one has to bear it only by humour, even if it is black humour.) But the turmoil was far from over.

In 1989, MAN B&W tried to purchase all shares of MBS Dieselmotoren, but were first blocked by the German cartel office and, in January 1990, by the German Ministry of Economics. Then, on 15 November 1990, a consortium composed principally of the German shipbuilder Bremer Vulkan and the Italian shipbuilding conglomerate Fincantieri Cantieri Navali Italiana, purchased 84 per cent of the Sulzer Diesel shares, with the remainder split between the former parent, Sulzer AG (10%) and Sulzer Diesel management (6%). The Phoenix that rose out of the flames was New Sulzer Diesel (NSD).

In January 1996 Bremer Vulkan and Fincantieri bought the outstanding shares and each owned 50 per cent of New Sulzer Diesel. But, soon a curious and tragic event befell the company. Shortly after the sale, Bremer Vulkan, which had been struggling for some time with heavy losses from its shipbuilding operations owing to heavy competition from lower-priced Asian products, found those losses overwhelming. It sought salvation from government and private sources, but

Gerard Bally and Peter Sulzer sign the papers creating MBS Dieselmotoren AG.

Peter Sulzer, his father, Georg, and Mrs Bally celebrate the birth of New Sulzer Diesel.

unfortunately no saviour emerged to bail them out. In February 1996 Bremer Vulkan filed for bankruptcy.

When Bremer Vulkan went bankrupt, its 50 per cent share was briefly picked up by Fincantieri, but the drama continued. Henry Schneider was a member of the New Sulzer Diesel management at the time, and a shareholder (he left the company in 1996). He recalls that during the first days of 1996 "a disaster started when Bremer Vulkan collapsed." Nobody was sure who would take over its shares of NSD.

The Finnish, Italian and Swiss cultures formally merge at the official birth of Wärtsilä NSD Corporation.

After some harrowing days in "no-man's land," in September 1996 it was agreed that all shares of NSD were to be transferred from Fincantieri to the Finnish conglomerate Metra Corporation, the parent of another engine giant, Wärtsilä Diesel. The new company thus formed was named Wärtsilä NSD Corporation, and it happily received the blessings of European anti-trust authorities early in 1997. Finally, in April 1997, Metra held 87.6 per cent of the new company's shares with the remainder held by Fincantieri.

To some observers, the merger was a curious one at best. After all, Sulzer had been making diesel engines for 100 years, Wärtsilä for only 40. In fact, in 1953 Wärtsilä had become a Sulzer licensee, and subsequently built many Sulzer-designed engines for installation in ferries, cruise ships, and icebreakers.

In an earlier joint project with Sulzer, in fact, Wärtsilä's Finnish shipbuilding division had built 19 Russian icebreaking cargo vessels, to run the 6,000 treacherous miles between Vladivostok and Murmansk. The ships' engines were configured to directly drive each ship's propeller for open-water sailing. But when the ships engaged in heavy ice navigation, the engines could be hydraulically coupled to the propeller to prevent them from over-torquing and stalling if the propeller became hardbound in thick ice.

To the Sulzer brand's benefit, the merger gave it a clear mandate to concentrate on two-stroke engines, less so on four-stroke engines, which are Wärtsilä's specialty. But, as nearly all mergers do, this one caused some rather awkward chemistry between stalwart Swiss employees and the new Finnish culture; several key managers left the company, including Dr George Lustgarten, director of technology.

The merger had the immediate advantage of producing a larger four-stroke engine product line and a broader service network for the two former competitors. Prior to the merger, New Sulzer Diesel had only nine network companies to perform sales and service in the field, but employed many agents around the world. Now, the merger into Wärtsilä NSD Corporation has created a force of 46 network companies, some of them originally set up by Wärtsilä in developing nations that have continually growing demand for stationary power plants, for which the four-stroke engine is ideal.

Igarka, a Russian icebreaking freighter with Wärtsilä-built Sulzer four-stroke engines. Below, a 16-cylinder ZA40S built in Mantes, France.

Licensing today

When Sulzer Brothers' Diesel Department (now Wärtsilä NSD Switzerland) ceased manufacturing in Winterthur, in 1988, it became a design, engineering, and marketing company, selling technology to shipyards and to shipowners through its licensees. As a result, the terms of its licensing agreements are quite demo-cratic; the document is, in the friendliest sense, a "know-how" license — a transfer of knowledge, technology, and experience, for the benefit of both parties. Some of that technology is patented, some of it not (Sulzer, New Sulzer Diesel, and now Wärtsilä NSD Switzerland hold more than 500 patents in about 150 areas of diesel-related technology, most of them pre-dating the merger).

Sulzer brand engines are manufactured today by more than 20 licensees in Europe, Asia, and the Americas, with the greatest production concentrated in China, Italy, Japan, Korea, and Poland. The objective of the licensing agreement is a long-term relationship, freely flowing with know-how and up-to-date technology, aimed at producing reliable, efficient, fairly priced, uniform, easily installed and environmentally friendly engines. It's a lot to ask, but there is mutual benefit for all: the licensee produces a saleable (presumably profitable) product; Sulzer builds a pool of R&D capability to improve that product, and achieves presence in key shipbuilding countries.

A licensing contract runs for 10 to 15 years, though it may be revised at mid-contract. The contract defines the licensee's territory, and indicates the types of engines it will be expected to build. It commits Winterthur to providing the necessary drawings, materials lists, standards, operating and maintenance manuals, and the complete training to build those engines. The contract also commits Winterthur to

A design team prepares drawings and specifications for licensee application.

advise and train licensees on operation, testing, and installation of the engines they build.

Naturally, the contract spells out payments to be made by the licensee, in return for this technology transfer. First, a down payment is made at the start of the relationship. And a documentation handling fee on drawings is periodically paid. (There is no charge for the drawings themselves, a fair deal, as a typical engine needs some 4,000 drawings. Winterthur supplies 150,000 to 200,000 drawings to its licensees each year; considerably more when a new engine is introduced. Drawings, once on paper or Mylar, are now on microfilm or digitised in CAD software.)

The licensee is also expected to pay a royalty against sales of its spare parts, based on their selling price, as this is a source of considerable revenue for both parties.

The lion's share of the licensee royalty is against sales of complete engines. Here the royalty is paid not as a percentage of the selling price, but as an amount per horsepower sold. (Currently, the Sulzer brand order intake worldwide averages some four to six million horsepower per year, so the royalty is substantial enough to support the research and design activities that are the Sulzer brand's life blood.) These industry-wide practices assure the home office in Winterthur a standard rate regardless of the licensee's selling price, which protects it against market price fluctuations owing to pressure on licensees to remain competitive. (And it saves Wärtsilä NSD Switzerland accountants the discomfort of ever having to look into licensees' books.)

All payments are in Swiss Francs. But there have been some rare, if earnest, attempts at making exceptions to that rule: In the late 1970s, an American delegation from the Westinghouse company wanted to pay

A 12-cylinder RTA84C engine is tested at Japan's Diesel United, one of the Sulzer brand's prime licensees.

its royalties in crispy American "greenbacks." The Swiss were, shall we say, disinclined to accept. The Americans boldly challenged them to a bowling match, the winners to choose the contractual currency. But the Yanks did not realise that Swiss-style nine-pin bowling is quite different from American-style ten-pin. Americans have been paying in Swiss Francs ever since.

Of course, contract terms are standardised as much as possible. But there are occasional exceptions to the standard owing to international law and other local exigencies. With licensees now the sole manufacturers of the brand, there have also developed some sharp "cultural" changes in the marketing of Sulzer brand engines. For one thing, in by-gone days a shipowner might come directly to Winterthur, have dinner with some very familiar managers and, after a snifter or two of cognac and a couple of deep draughts of a cigar, sign a contract for several engines on the back of a napkin.

Now (under the contemporary influence of the ubiquitous legal profession), licensee contracts are 30 pages long and growing, and the relatively few engines traded by Winterthur are handled by a small specialised group called the Marine Engine Sales Department.

Also, in a world with a limited number of diesel works capable of building these magnificent machines, many Sulzer-brand licensees also manufacture engines for the main competition, MAN B&W. And, to make matters even more intriguing, one of them, Mitsubishi Heavy Industries, produces engines under its own brand name.

Multiple licensing places enormous competitive pressure on the design and marketing strategies of the main brands, and increases the delicacy of relationships among all parties. After all, if a licensee possesses the drawings, specifications and know-how to produce one engine brand, and is producing the competitor's brand at the same time and on the same shop floor, the border between the two competitors' technologies may be easy to cross, or just difficult to find. So a high degree of mutual trust is necessary for the licensing arrangement to work at all.

There is another significant cultural change that came with the maturity of so many licensees. No longer can the main, European-based creators of diesel engines play "king" to their licensees "servant." The two parties are more and more on an equal footing, with Winterthur persuading rather than dictating to its partners,

An overview of the diesel engine assembly hall of the Sulzer licensee Cegielski, in Posnan, Poland.

particularly in the case of manufacturers in East Asia.

Japanese licensees gained technological pre-eminence from the tremendous boost of their post-World War II industrial recovery. By carefully looking into products and methods, Japan mastered the arts of pre-assembly and quality control in shipbuilding, which it now applies to engine manufacture. Licensees such as Diesel United and Mitsubishi, but more and more Korean firms, such as Hyundai and Hanjung, do research to improve their licensed engines, and pass that technology back to Winterthur, which results in general improvement of the product worldwide. Similarly, the Polish licensee, Cegielski, has extended manufacturing experience, and is geographically close enough to Switzerland, to have developed a strong spirit of technological cooperation.

This changing culture is perhaps even more sharply reflected in China, which has undergone great change in the past three decades. The first licensing agreement with China was signed in 1965, just before the onset of the Cultural Revolution. Sulzer Brothers licensed the Chinese to build one product, a 760-mm bore engine

The smallest Sulzer two-stroke engine, a 5RTA38, is run in the test bed in Shanghai, China.

of 12,000 hp. But once the revolution started, there was no communication between the parties. By 1976 matters improved. The Chinese began to build ships, and at first they ordered their engines from other Sulzer licensees in brotherly communist countries, such as Yugoslavia and Poland. Then they began to build their own engines, a mix of Sulzer and MAN, which were not terribly successful owing to the difficulty of their supplying spares. It was then that they approached Sulzer for a new license agreement.

The first large delegations from China came to visit Sulzer at the end of 1978, after the lapsed first contract had been renewed. They were experts eager to learn, and to see Switzerland. But they lived fairly controlled lives: they carried no money, except for their leader. They stayed privately at the Chinese embassy, not public hotels. They didn't mingle with their Swiss counterparts. Although English was the *lingua franca* of business and friendship, there was always the risk of misunderstandings. When the Sulzer delegation reciprocated with visits to China, meetings were formal and the guests were always accompanied by Chinese "colleagues." The Swiss and Chinese travelled as separate groups, took separate flights. To help them get an economic start, Sulzer's engineers proposed that the Chinese be conservative and buy second-hand machinery; though some of their workers couldn't yet read a vernier scale, they insisted on buying new machines to maintain their pride. They had no capable foundries, so the Swiss taught them how to cast.

Today, the facade of China has changed considerably. When the original 1978 contract was signed all members of the top Chinese delegation were over 65 years old; when the contract was renewed a few years ago the oldest was 39, the youngest 27. Sulzer and its Chinese licensees experienced the normal growing pains of a new relationship. Sulzer wanted quality, the licensee wanted production. But, because the Chinese understand Swiss humour quite well, everybody worked it out.

And the level of comfort between the two parties rose dramatically, even as the formality fell away. Delegations can now meet on an open and friendly basis, and enjoy a good business dinner together.

A similarly warm and quite remarkable change has taken place with a Japanese licensee, Diesel United. The Japanese have a wonderful ceremony to bless a new engine before it is sent off to its waiting shipyard. Everyone concerned with the project — from top managers to the workshop labourers — dons clean coveralls and shiny hard-hats, and joins in a blessing of the engine. The purpose of the ceremony is to go around the engine with a cup of *sake* in hand, and anoint every component that could possibly cause problems (this can be revealing). In the early days, the ceremony was exclusively Japanese — no Europeans joined in. Now the Sulzer engineering staff working with Diesel United are invited to be part of the ceremony.

With such a mix of licensees, such a wide range of cultures, maturities, and with varying manufacturing methods, the dilemma for Winterthur always is: for whom do we design engines? Should it be for the newest or the most mature member of the family? Should we aim for high material cost and low labour cost? Or

The Winterthur team joins Diesel United in the blessing of a new engine with cups of *sake*.

the other way around? This is a most vexing issue, as the cost of an engine varies widely from licensee to licensee.

Sourcing of components for the manufacture of engines and their peripheral systems is another vexing area. Ideally, parts coming from suppliers should be identical, manufactured to the same standard by all suppliers, so that all engines have interchangeable parts. The stricter the adherence to these standards, the more customers enjoy full interchangeability. But, in order to save costs and hold their competitive edge, most licensees want some flexibility in their cost and pricing structures, because price has become the most compelling consideration in choosing a main engine.

So, in the end, a license agreement for Sulzer diesel engines is not just a legally binding contract; it is an expression of mutual trust, of healthy confidence between Winterthur and the licensee. Despite what sometimes seems to be the brutal drive among licensors toward technological nirvana and competitive purgatory, it is on that basis that nearly all the world's large diesel engines are made and sold. It is a satisfying basis, even with its risks, on which to do global business.

(This very same basis, by the way — a handshake, a smile, a shared meal — is the very way in which the Sulzer diesel has established respect among shipowners as well. Peter Sulzer is the Executive Director of Wärtsilä NSD Corp., Chairman of Wärtsilä NSD Switzerland, and the last descendant of the Sulzer family still engaged in hands-on marketing of diesel engines. He is also the elegant, charming, multilingual, and unofficial roving ambassador for Wärtsilä NSD. He recalls with pride his having known Chairman Y.K. Chang of the giant container shipping company, Evergreen. They met in 1974, in Taipei, when Chang had only a handful of ships. Now he has a fleet approaching 100 ships, and is the largest independent containership owner in world. Peter Sulzer visits the Evergreen offices every year, just to see his old friend. Despite the clamour of the competition from northern Europe, for 15 years nearly every Evergreen ship has had a Sulzer diesel engine.)

Peter Sulzer joins Chairman Inaba of Ishikawajima-Harima Heavy Industries in a gesture of hands across the seas.

Ever Dainty, one of the fleet of hard working Evergreen containerships that is driven by a Sulzer diesel engine.

License Care

The main difficulty in the licensing arrangement is that there can be only one engine that bears the name "Sulzer"; no matter where it is built customers expect one quality standard. In order to watch over the licensee relationship, therefore, Winterthur has a dedicated department it calls License Care, with some 20 technical specialists whose full-time assignment is to look after the needs of the licensees.

The License Care and Application Engineering Department is headed by Beat Wiederkehr, whose heart "beats" for the Sulzer diesel engine. The department's oddly genteel name was chosen out of respect for the genuine cultural sensitivities that exist in the relationship between this small Swiss entrepreneur and its many diverse international licensees (particularly those "tigers" of the Pacific Rim). It is well known that most nations, no less Asian nations, have enduring pride in their history and culture; they want to be partners in a world where Eurocentrism is a thing of the past.

So, New Sulzer Diesel changed the original name of its License *Support* department to that of License *Care*, as a subtle linguistic indicator that licensees do not need Winterthur's *support* so much as they warrant its *care*.

The License Care team, (headed by Wilfried Pernter, a gentleman with style and strength) is charged with helping licensees produce a trouble-free engine with one high level of quality, by assuring a constant multi-directional flow of information, constant communication to solve problems, and the provision of up-to-date materials (such as the three-dimensional piping drawings that enhance CAD engine-room design, developed by Jakob Jäggli's CAD team). The License Care team assists each licensee to calculate the costs of producing an engine, taking into consideration the

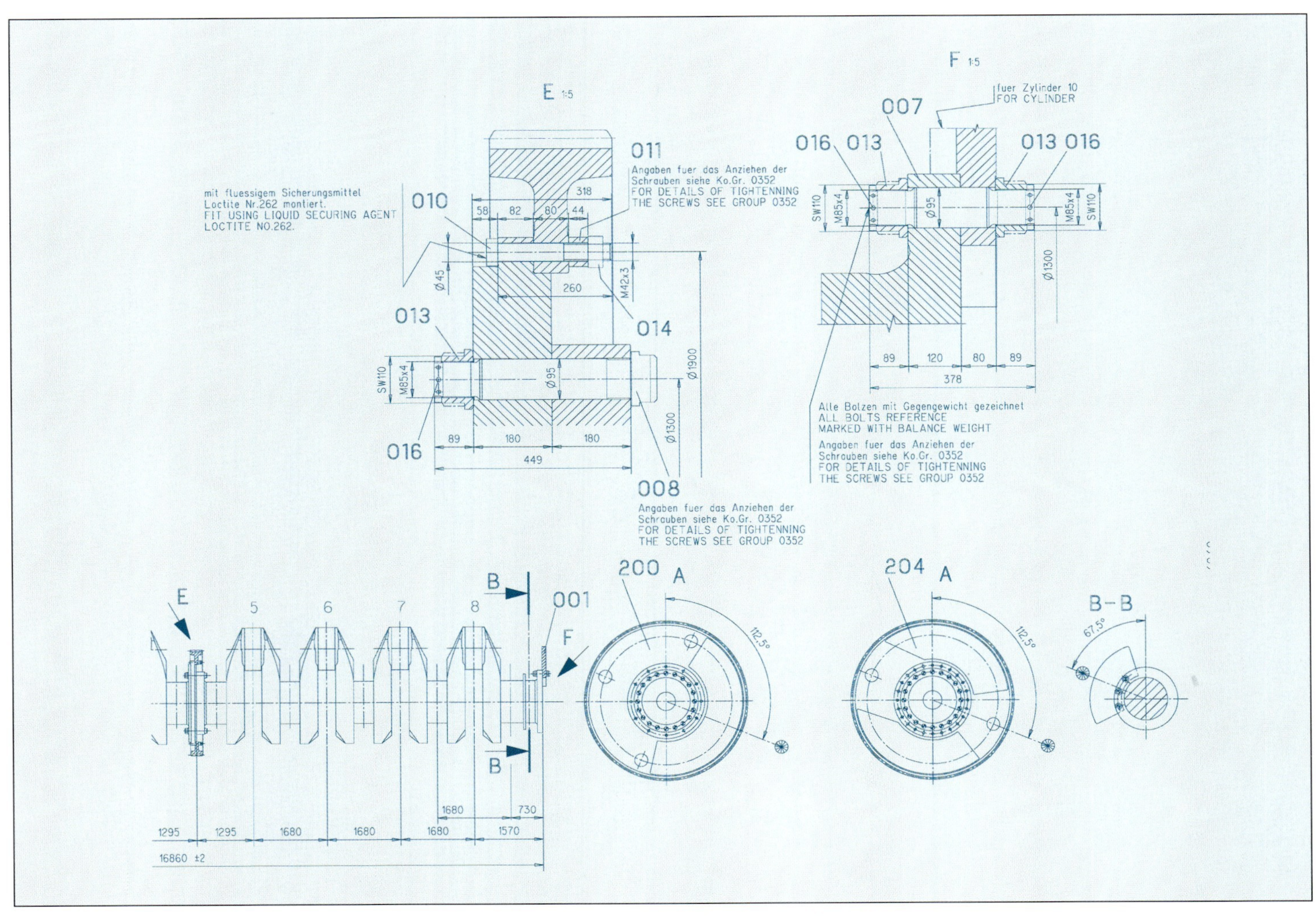

A typical engineering drawing, giving details for securing a crankshaft assembly.

differences in their equipment, methods, experience, and culture. A high degree of diplomacy and sensitivity are required to do this work.

Under license agreements, the licensee is fully responsible for every engine's meeting Sulzer specifications. But, as can easily be seen, there are sometimes grey areas when an engine does not live up to those specifications — when the problem has been caused by either a design fault or a manufacturing fault, which must be resolved amicably by the License Care team so that the licensee is not harmed and shipowner is best served.

Further, each new engine type must undergo a "Type Test" required by classification societies. And there are the equally stringent international rules of ISO 9001, which govern so many aspects of engine quality. (License Care Quality Assurance people monitor the ISO 9001 standards, so that quality failings can be identified and traced back to their source for rectification and future application. There are no insurance policies against manufacturing failures, and no matter the cause, if engine quality is less than excellent, the brand name suffers.)

Nowadays there is enormous pressure from shipowners for the shipyards (and engine manufacturers) to extend the guarantee of a newly developed engine. While it has been tradition for such shipyard guarantees to last one year, market pressure is demanding longer guarantees on components and even entire engines. That, again, exerts enormous pressure on the licensees, which can be relieved by the Care department.

As part of that care, Winterthur invites representatives of all its licensees to Switzerland, to discuss the highlights (and lowlights) in engine quality, design and manufacturing. A simple gesture, such as Sulzer diesel people inviting visitors to their homes to entertain them, goes a long way toward solidifying relations. There is no better way to solve a problem than breaking bread and drinking wine, face-to-face, with a colleague who builds that engine 10,000 miles away.

Licensing Today

In the last decades of the 20th century, many of Sulzer's licenses (and licensees) expired, in such diverse locations as Hungary, Moravia, the United Kingdom, and Belgium. In the UK, some shipyards produced their own engines; they are gone. The Sulzer brand now has 21 licensees with more than 30 diesel works in China, Croatia, Finland, France, Germany, Italy, Japan, Korea, Poland, Spain, Turkey, and USA.

Yet, even with the demise of so many engine manufacturers, there is nonetheless a worldwide engine production overcapacity (some surveys estimate that the capacity is double the industry's needs). As a result, there is a continuing price race among large-engine builders (as there is among large shipyards), and the competition is often irrational.

Hipolit Cegielski (1813–1868).

Some of Sulzer's older licensees have wonderful, often surprising, histories that give them a special relationship with Winterthur and its own grand traditions:

In a history almost as long as Sulzer's, the licensee H. Cegielski-Poznan has its origins in 1846. That was when Hipolit Cegielski, a doctor of philosophy and teacher, established a hardware store in Poznan; it soon became a small factory producing ploughs and other farm implements. From that humble start, Cegielski eventually grew into a prime manufacturer of steam and diesel railway locomotives and machine tools. Cegielski became a Sulzer licensee in 1956 (and a Burmeister & Wain licensee in 1959). It has manufactured more than 1,200 low-speed marine propulsion diesels up to 760-mm bore, and 6,000 medium-speed auxiliaries, working closely with Polish shipyards.

Mitsubishi Heavy Industries was started only 50 years after Sulzer, in 1884, by Yataro Iwasaki, who leased a government-owned shipyard, renamed it Nagasaki Shipyard & Machinery Works, and started building ships. In early 1925 Mitsubishi became a licensee of the Sulzer Brothers, and has remained a Sulzer licensee since. After several incarnations, in 1934 the shipyard became Mitsubishi Heavy Industries, the largest private firm in Japan, manufacturing ships, heavy machinery, aircraft, and railway cars. After World War II, Mitsubishi was dismantled by the Allies into three separate entities, which were rejoined in 1964 as today's international conglomerate, Mitsubishi Heavy Industries. MHI produces a large number of Sulzer brand diesel engines. They typically build Sulzer's larger engines, such as RTA84C, but recently started production of the RTA48T.

Hyundai Heavy Industries, of Ulsan, Korea, has been a Sulzer licensee since 1975. It is the largest shipbuilder in the world. And with the 100 to 120 two-stroke engines it builds each year, totalling on average more than two million horsepower, it is the largest engine builder in the world as well. At the end of 1997, Hyundai reached a grand milestone: It surpassed the aggregate of 20-million horsepower by delivering a turnkey power plant to an Indian client. The astonishing thing about that accomplishment is that the company started only in 1979, so it accumulated that output in less than 20 years. (Unfortunately today, the overwhelming majority of its two-stroke engine production is of the "other" brand.)

China has a half dozen engine-building licensees, whose several diesel works are located in Dalian, Chongqing, Yichang, and Shanghai. But only one company administers the Chinese licensing agreement: the central government's China State Shipbuilding Corporation. The first licensing agreement between Sulzer and China, signed in 1965, was never implemented; a second contract was signed in 1978. Chinese licensees have focused mainly on medium-size diesels, and

Peter Sulzer pays an "ambassadorial" visit to the Dalian, China, licensee in October 1997.

since 1997 they also produce the very modern RTA48T and RTA58T engines.

Diesel United, one of the few exclusive licensees to the Sulzer brand, has its headquarters and main works in Aioi, Japan, with branches in Tokyo and Kobe. Diesel United was established in 1988 by the joining of the diesel divisions of two Japanese industrial giants, Ishikawajima-Harima Heavy Industries and Sumitomo Heavy Industries. Their experience, however, goes back to 1950, when the first modern Sulzer engine was made in Japan. (Sulzer licensing in Japan, however, dates back to 1917, with an agreement signed with the Imperial Japanese Navy.) Including its predecessors, Diesel United has produced 35 million diesel horsepower. Today, Diesel United contributes considerable research and development to the betterment of the Sulzer diesel brand. It is the manufacturer that typically builds Sulzer's largest engines. Diesel United was the first to build the 66,120-hp 12RTA84C, and, as we shall see, has since built the world's most powerful diesel engine, the 89,640-hp 12RTA96C. It was also first to build Sulzer's smallest modern diesel, the RTA48T.

East meets West at Diesel United, in Aioi, Japan.

The RTA engine

The crisis of the 1970s, brought on by the closing of the Suez Canal and the oil crisis of 1973, was a shocking encounter with reality for the industrialised world and its dependence upon the free flow of oil. For Sulzer it was a worse encounter. When the crisis struck, the company had just completed development of its RND-M, an engine designed according to a set of priorities: the *highest* being reliability. That alone would not have caused pain and suffering to Sulzer, had it not been for the fact that one of the design's *lower* priorities was, yes, fuel economy. The engine used the firm's standard loop-scavenging exhaust system, with scavenge air and exhaust ports at the bottom of the cylinder and no exhaust valve in the cylinder cover.

That proved to be a serious problem, because ships were generally running at slower, more economical, speeds and the new breed of vessels with very large, slow-turning propellers, needed engines of very long stroke. Loop scavenging was fine for shorter stroke/bore ratios (say, up to two), but beyond that cylinders were too long for effective scavenging. The RND-M engines were plainly out of tune with the market's sudden, frightful, shift.

Sulzer researchers set to work developing a new series of longer stroke engines that would be able to run at lower

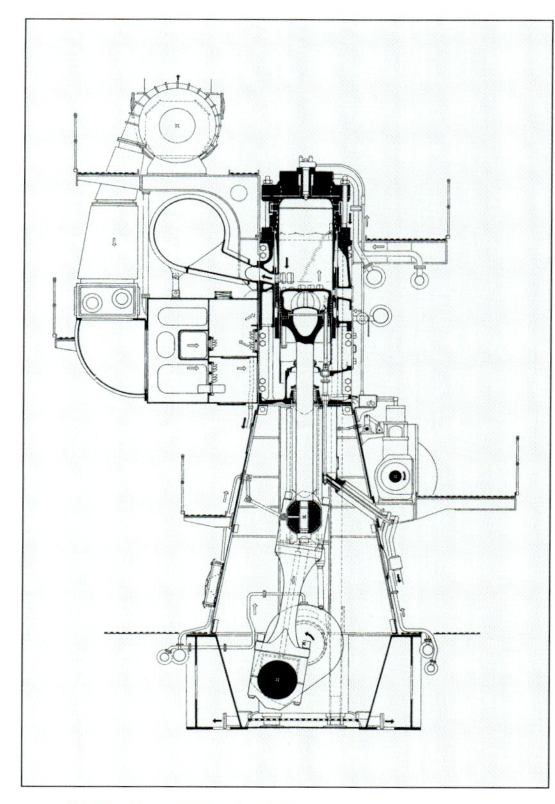

RND-M engine (1973).

revolutions and therefore have higher efficiencies. The new engines, the RL series, were specifically tailored to the slow-steaming, high-economy ships that were fast becoming the rule of the day. The RL engines were also the first to be equipped with variable-injection timing, to maintain ideal combustion pressures through all the load range.

But longer stroke engines were not alone a sufficent solution. After further development, in January 1982 New Sulzer Diesel announced a major shift in its design technology, driven by the longer strokes of its new higher-powered engines: it changed its scavenging from "loop" to "uniflow". (The shift to uniflow scavenging for New Sulzer Diesel was so fundamental and powerful that in retrospect it appeared more like a conversion of religion than a mere shift in science.)

Uniflow scavenging, however, as should be obvious, was more complicated in that it required the addition of the exhaust valve and its timing system — and exhaust valve technology of the day didn't promise long enough valve life.

However, the exhaust valve in Sulzer diesel engines proved to be extremely reliable, with time between overhauls sometimes exceeding 25,000 hours. The new series of engines was designated "RTA". The RTA series virtually exploded onto the market, with the announcement of six models with bores ranging from 380 to 840 mm, and stroke/bore ratios of 2.9. And, as a result of the immediate market acceptance of the design, more models followed in fairly rapid succession. Three years later, in 1984, two new bore sizes were added to the original family, and the largest engine, the RTA84, was released in an "M" version — all three had an increased stroke/bore ratio of 3.45, as did the RTA72 added in 1986. In 1988 the

The 4,038-TEU P&O containership *Shenzhen Bay*, built by Ishikawajima-Harima, is driven under a full load by a Sulzer 9RTA84C engine.

RTA84 was again modified and uprated to a "C" version, which was designed specifically for a very significant market — large containerships in transoceanic liner service (the "C" meant Containership). Six years later the company was to announce its second, larger "C" engine, the RTA96C (about which much more will be revealed later).

The 1990s were the beginning of a new direction for the entire RTA catalogue, as more and more engines were carefully tailored to specific shiptypes. Thus, in 1992 the original RTA52, RTA62, and RTA72 engines of 1984 were upgraded, designated "U" (for Upgrading), and targeted for product tankers, medium-size containerships, reefers, and general cargo ships. In 1992, a special VLCC prime mover was introduced called RTA84T (the "T" of course stood for Tanker). It had special features such as load-dependent liner cooling and variable exhaust valve closing for the special sailing conditions of VLCCs. In 1995, new "workhorses" with stroke/bore ratios of 4.17 (RTA48T and RTA58T) were earmarked for smaller, slower-moving ships, such as Handysize to Capesize bulkers and tankers. In 1996 their larger sibling, the RTA68T, with stroke/bore of 4.0, joined the family. Finally, in 1997, the three "RTA-U" series engines were all redesigned with more compact dimensions and reduced manufacturing costs, and slightly higher power through higher effective pressures and piston speeds, and designated "B" versions. The Sulzer RTA series today is still divided broadly into these three types of engines, generally by their stroke/bore ratios: RTA-C with stroke/bore of about 3; RTA-U engines around 3.5; RTA-T series 4.0 or higher.

The conception for the RTA engine was an all-consuming task for both the development team headed by Heinrich

A sampling of the typical reefers, tankers, bulkers, and VLCCs for which the RTA series engines are ideally tailor made.

Brunner and the construction team headed by Ulrich Vetter. A good deal of thermodynamic and tribologic research was done by the research department in configuring the engines to minimise wear.

The RTA84T was a particularly challenging project for the thermodynamic and mechanical engineers of Winterthur. The engine has the longest absolute stroke of any diesel engine, at 3.15 metres. (It could be longer, but that is the longest stroke that can be handled today by suppliers of crankshafts.)

In developing the RTA84T, however, researchers underestimated the influence of the long stroke and the low speed of 74 rpm. This combination not only produced uneven liner temperatures over the cylinder length, it allowed longer time in each stroke for combustion-created sulphuric acid to penetrate the lubricating oil and attack the liners. So, the cooling jacket design had to be completely revised.

The somewhat arbitrary division of engine types, and the heightening competition, led Wärtsilä NSD Switzerland, in 1997, to create three new Product Manager positions within its Technology Department, each to guide a specific family of engines to the market. The Product Managers' task, according to their leading light, Kaspar Aeberli, Director of Technology, is to spend day and night thinking about their engines, to act as links between design, manufacturing, and marketing, and to make certain that both mature products and new engines meet the requirements of the marketplace. This is no easy task, but it is a vital one to a company that, in the past, has occasionally lost sight of its market.

Each product manager determines what engines may be needed by first looking at market studies and shipowner

needs, then interfacing with the design and marketing departments to meet those needs. If, say, a problem develops in the field with an engine in his domain, it is the product manger's task to help decide whether future engines of the type will require modification, or that a new engine is needed.

To illustrate that point, in early 1998 Wärtsilä NSD Switzerland was in negotiations to obtain the contract for the engines for a series of ships for Dole, the American mass marketer of fruits and vegetables. The ships Dole proposed were 1,000-FEU reefers set to ply the banana trade between Costa Rica and the US mainland. Reefer ships, in addition to needing very reliable engines, draw a good deal of extra electric current for their refrigerating units, so they need plenty of safe, redundant power generation.

And Dole is conservative in its ship operations, and insists on turbocharger redundancy as well. This means a total power plant — engines and auxiliaries — that can become expensive. The product manager, Jürg Flach, along with his marketing colleagues, had to juggle various considerations, including a shaft generator for added power, or a package deal that included auxiliaries. Ultimately, Dole chose the 8RTA72U. The engines were sold through the Marine Engine Sales department in Winterthur and will be built in Korea.

The product managers are obviously at the forefront of the race for market share, which can be brutal. There is pressure on engine builders to garner market share by increasing the length of engine guarantees, as well as reducing prices. So, in addition to their technical prowess, the Product Managers must have marketing "smarts" as well. The managers, who make all this possible are Rudolf Demmerle (for the "C" engines), Markus Geist (for the "T"), and Jürg Flach ("U").

Despite their keen attention to the market, licensees, and customers, and their obvious qualifications for their tasks, there was mild resistance to the product management concept at first. But, it has become clear to all that these young managers (at least young-minded, as Demmerle is quite mature compared to the others) bring a fresh outlook and added value to their products.

They are involved in the market, and they offer owners an opportunity to discuss each project with a single technician who knows the engines intimately. As the vice president for

Towing-tank tests of the Dole reefer hull were run at the Hamburg Ship Model Basin.

marketing, Christoph Studinka, feels, "These guys are very welcome, because their overview of the needs of the market and the licensees can only result in a better product."

The RTA engine has become the centrepiece of the Sulzer two-stroke range, and in its various incarnations has driven engine research into fresh new fields of creativity. RTA engines, for example, were the first to exceed 50 per cent thermal efficiency; they were the first uniflow-scavenged engines to extend time between exhaust valve overhauls to 30,000 hours, or four years running time; they included the first "superlong" stroke engines. At the dawn of 1998 (the 100th anniversary year of Sulzer's running its first diesel engine), there were more than 1,960 Sulzer RTA engines in service or on order, totalling more than 40 million horsepower, built by many licensees. The Sulzer RTA diesel engine is truly the workhorse of the world.

Perhaps it is best to summarise the quiet but vital contribution the engine has made to that world by quoting Markus Geist, who ought to know: "You never think when you buy a banana, that it was delivered to market by a ship driven by a Sulzer RTA engine."

Tailor made engines

In the 1970s, despite heavy competition, the Sulzer brand enjoyed the number one world market share in marine diesel engines. But when the takeover by the German company MAN of the Danish Burmeister & Wain took place in 1980 it automatically created a company with a higher market share than Sulzer, enhanced by the fact that it offered customers a certain amount of security of size. Though Sulzer's share of market for large-bore, low-speed containership engines remained about 50 per cent, it proved relatively weak (with market share varying around 20 to 30 per cent) in engines for those feeders, tankers, and bulkers mentioned above, which have become known as "standard ships".

The term standard ship is applied to a range of very popular tankers and bulkers whose capacities generally fit within a common range, and which are similar regardless of the shipyard that builds them or the shipping company that owns them. Among tankers, the Aframax (average capacity of 100,000 dwt) conforms to a freight index for tankers trading to the Americas; its larger brother, the Suezmax (average 140,000 dwt) carries the

The bulker *Romandie* steams along powerfully by virtue of her 5RTA62U engine.

The Norwegian VLCC *Berge Stadt* pushes aside plenty of water propelled by her Sulzer RTA84M diesel.

maximum full-load draught permitted in the Suez Canal. Among bulkers, there is the small Handymax (average 45,000 dwt), the medium-size Panamax (average 70,000 dwt and limited to the 32.2-metre breadth for passage through the Panama Canal), and the large Capesize (nearly 160,000 dwt average).

These ships have a typical range of power requirements as well, but in the past not every ship had an engine to meet those requirements. So, as the RTA series became established, New Sulzer Diesel applied the term "tailor-made" to the design of these specialised engines.

Ideally, any type of ship ought to have the lowest-powered (therefore the least expensive) engine required to satisfy its speed requirements (typically, the Handymax needs an engine of around 11,000 hp; Panamax about 13,500 hp; Aframax about 16,500 hp; Suezmax and Capesize around 22,500 hp). The tailor-made engine makes individual ship operation more economical. And standardising engines for each ship type helps licensees to optimise the manufacturing methods, and thereby reduce production costs.

There are many factors that must be considered in the design of a new diesel engine, including the ultimate user, the shipowner. The question then is: For whom and for what do you design a new engine?

***Hanjin Osaka* plies her container trade taking her power from a Sulzer 10RTA84C diesel.**

Felix Prochaska can surely answer the question. Prochaska, who joined Sulzer Brothers in the mid-1960s, is a trained naval architect, mechanical engineer, creative thinker, and Manager of Marine Projects and Marketing (though he's a technical man, the marketing responsibility was added to his job description because of the compelling need for a friendly interface between the two sides of the business). Indeed, it is the marketing side of Marine Projects Department that must first attract knowledge of new ship projects to enhance the prospect of engine sales. This is largely accomplished by regular contact between Winterthur, its licensees and network companies, who are most likely to first learn of new projects. Then, when an engine is to be chosen for a new ship, the department sends the owner a detailed questionnaire, or sets up meetings for more detailed discussion. From the answers to the questionnaire, designers do preliminary power/speed analysis, producing a set of curves predicting the power required to attain various ship speeds. Then, using computer software developed by the

The propeller form is fundamental to the choice of engine for the ship.

MARIN towing-tank facility, in Wageningen, Holland, designers define the engine parameters (bore, stroke, rpm, number of cylinders, etc.) to suit the ship.

This process, however, may suggest more than one engine being suitable to the ship.These variations are weighed by the shipowner as to efficiency and cost to make the selection. (As an example, say that a new ship could be served by engines of five, six, or seven cylinders of differing bore sizes but comparable power. The five-cylinder engine may be the least expensive, but it likely produces higher torque variation and vibration, and requires second-order balancing, torsion vibration dampers, and side stays. The seven-cylinder engine has minimal vibration, but higher initial, operating and maintenance costs. The six-cylinder engine, though falling between them, still might or might not be the ideal choice.)

Then there is the propeller to be considered. According to Felix Prochaska, the governing factor in the choice of a ship's main engine is the velocity of water entering the propeller plane — in a sense, the ship's drive train is typically designed backwards, from propeller to engine, rather than the reverse. Most propellers today have four, five, or six blades. Although the relation between the number of cylinders and the number of propeller blades has no effect on vibration, the phasing of the propeller blades with the crank webs is critical and, if wrong, can be a disaster. The minimum ballast draught of the ship, as defined by the IMO, defines the maximum diameter to assure fully immersed propeller tips. Thus, the MARIN program helps in the selection of the ideal propeller, by seeking the largest diameter that will fit the ship, running at the lowest rpm, to obtain the highest efficiency, and lowest fuel consumption.

In summary, to find a suitable engine, you choose the smallest one possible to deliver the full power and speed required by its propeller. Or, in other words, you choose a tailor-made engine with a top power point close to the ship's maximum needs (including all margins for ageing, hull fouling, wind, and sea states, as well as operational and propeller light running margins). Then, the contracted maximum continuous rating of the engine is close to the rating the selected engine can deliver, so the shipowner does not have to pay for horsepower that the ship does not need.

Tribology

Among the most fundamental and absorbing problems in the design of high-powered, low-speed engines, is what is termed "piston running behaviour," or that old black magic, "tribology." The dictionary definition of tribology (from the Greek, τριβος, "a rubbing") is "the science of the mechanisms of friction, lubrication, and wear of interacting surfaces that are in relative motion." Suffice it to say that the cardinal tribological problems engine designers face are the "rubbing" of shafts in their bearings and, more important, between piston rings and cylinder liners. At their surfaces, these hard, overheated, metal masses inevitably produce friction and wear; it can not yet be totally avoided.

As power concentration, firing pressure, and mean effective pressure have increased, the result has been higher forces exerted on liners by the rings. This tends to squeeze the oil out of the fine space between ring and liner, and promotes higher potential wear. Tribologists are charged with deriving the mechanisms to disperse the pressure and keep the oil film right where it belongs, and in good condition, thereby limiting wear to an absolute minimum.

The industry is still on the learning curve of this challenge, seeking new materials, new machining methods, new lubricants, and generally cleaning up design- and manufacturing-related problems. This task has been, according to one Sulzer tribologic guru, "... part success, part catastrophe. You can work day and night to design pistons and cylinders for perfect running, but if at the end of the process the cylinder liner honing man has a bad day, you can destroy the reputation of a company."

It is a fundamental rule of thumb that the larger the engine the more intense the problems. With massive metal castings, larger dimensions and cumulative tolerances, bigger variation of geometry, everything adds up to the need in these dynamic machines for great control of sensitive, microscopic tribology. There will be a good deal more discussion here about research and testing that go into the conquest of tribology, as it is an integral part of the design of the engine of the future. But, perhaps this is a good time to look at one method by which tribology has been conquered in the everyday life of the brand — by the dutiful and precise construction of the Sulzer diesel engine.

The proper study of tribology is a greasy business indeed.

Engine construction

Many people may view the big diesel engine as a boring (no pun intended), heavy application of old technology and old engineering. But, as with its operational design, the diesel engine's construction has undergone decades of stepwise advance, modification, and modernisation, some of it quite incredible.

The earliest Sulzer marine diesel engines were logically built in much the same form (and by many of the same people) as their reciprocating steam engines: overly massive. Individual A-frames were made in cast iron, with integral cylinder jackets, and mounted on a single massive bedplate.

For the first two-stroke S-type engines of 1905, however, individual cylinder blocks were carried on column bolts, braced by diagonal tie bolts, with a single-piece cast-iron bedplate. In 1912 Sulzer made a more substantial marine crosshead engine, with cylinder jackets mounted on separate columns set on the bedplate. The engine, which also had interchangeable liners, was installed in the *Monte Penedo*, a German-built freighter of the Hamburg-Süd Line, and one of the earliest diesel-driven ocean-going ships. Long tie rods from the cylinder covers to the bedplate kept the engine in compression against the explosive tensile loads created by the repeated cylinder firings.

By 1917, Sulzer abandoned tie-rod construction in favour of a box-like structure of cast-iron columns and bedplate bolted together. These formed a rigid beam that withstood firing load tensions, as well as it resisted deformation transmitted from the flexing hull into the engine and crankshaft when a ship ran in a seaway. Over the next 20 years the design was refined in stages. Greater structural rigidity was achieved by bolting the individual A-frames together.

A cast-iron bedplate in the Winterthur shop ready for drilling (1912).

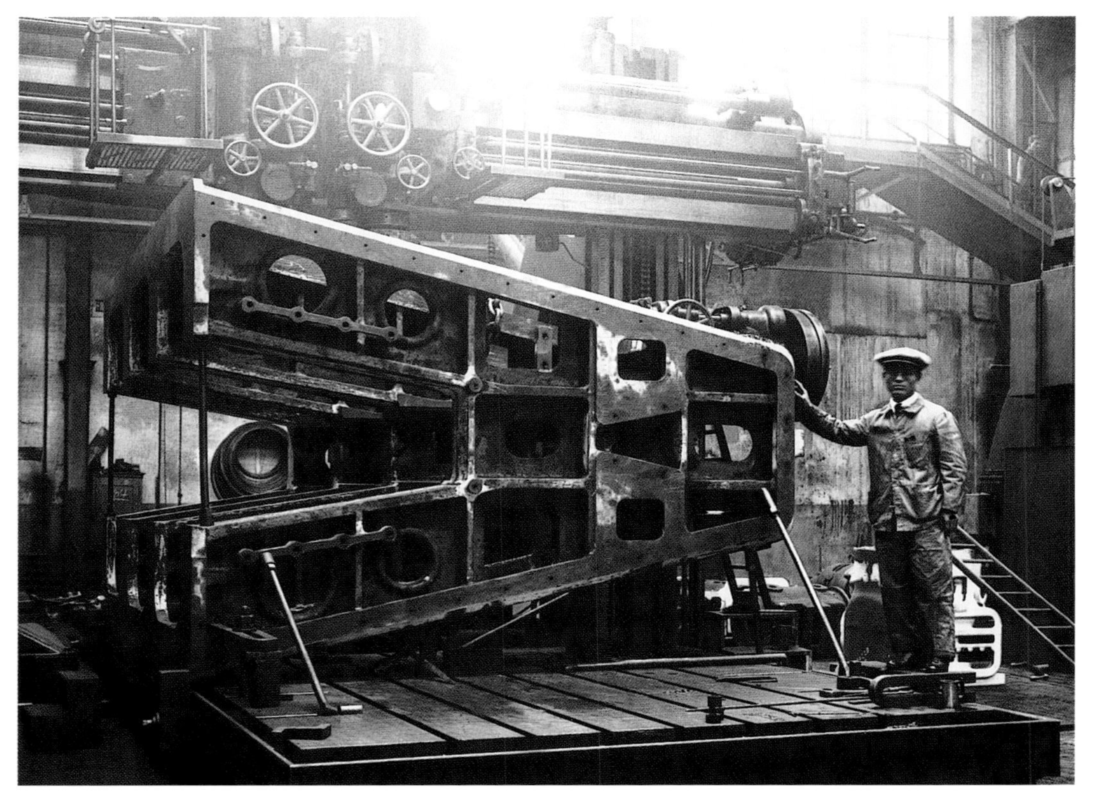

The frame of a three-cylinder S-type engine being machined (1925).

From 1941, cylinder blocks were also bolted together to give a continuous beam. In cases where light weight was desired, bedplates and columns were fabricated in cast steel. Attempts by Sulzer to lighten the structure further, however, using thinner castings and long tie rods running from atop of the block to beneath the bedplate, failed to produce rigid enough engines. So, around 1931, Sulzer reverted to cast-iron construction without tie rods, which was satisfactory to withstand maximum cylinder pressures that had risen to 50 to 55 bar. (The SD engines were Sulzer's last two-stroke crosshead engines fabricated with large iron castings. They were succeeded in 1950 by the RS types, with welded bedplates and columns.)

Welded construction of frames and bedplates had been employed as early as the 1930s

An ST68 being fabricated from cast columns and cylinder blocks (1926).

Cylinder covers and pistons stand before a 10-cylinder SD72 engine (c. 1940s).

by some European engine builders, such as Doxford, in England. But during World War II welding techniques necessarily made great advances owing to the Allies' intensive emergency war-production programme, in which welding largely replaced less-reliable riveting in the fabrication of warships and cargo vessels. After the war, ribbed engine castings were replaced by a box-like, double-walled welded structure for the bedplate and columns. This, together with the large modulus of elasticity of steel (approximately twice that of cast iron), made welded engines less susceptible to deformation and vibration, even though engine weights were reduced by about 15 per cent.

The first large Sulzer two-stroke engines of welded construction were introduced in the early 1950s. With their lighter foundations, these RS-type engines once again employed tie rods to hold their structures in compression against firing-related tension. Welded construction was further refined for the RD series, introduced in 1956, and continued through the development of the RND, RND-M, and RL engine series. With the RL series, Sulzer introduced a simpler bedplate construction with single longitudinal walls and an integral thrust block. This development culminated in the initial group of six engines in the RTA series.

Early two-stroke engines were short enough that their crankshaft could be forged in one piece. As bore sizes increased and cylinders were added, crankshafts had to be built up from individual crank webs and shrunk-in journal and crank pins, with keys to lock them in place. A semi-built form was ultimately adopted with two crank webs and their pin forged or cast as one unit and then shrunk onto the journals. As crankshafts increased

The complete crankshaft of the 10-cylinder ST68 engine being built for the Dutch twin-screw liner *Christiaan Huygens* (1928).

in length, combined with the cranks for scavenge pumps and blast air compressors, difficulties arose with torsional vibration. From 1912 onward, in order to master the vibration problem, methods were evolved to calculate the natural frequencies and additional stresses in complex shaft systems, which are today routinely handled by computer.

In the days before standardised SD type engines of 1941, there had been a tendency for each engine to be tailored to the specific wishes of the customers, in terms of bore, stroke, power and speed. Though the overall design was retained, there were often significant differences among them in details. This demanded a new rationalisation of engine construction, to reduce the number of design variations, and thereby obtain reasonably economical production. This was effectively the start of serial production of Sulzer two-stroke engines. All parts that were independent of the bore, stroke and number of cylinders were given a standard form. Such standardisation had far-reaching value in making components interchangeable, achieving economies of scale, reducing parts stocks, and speeding the delivery of engines and spares.

Connecting rods for early two-stroke engines (1929) compared to the running gear of an RLA90 engine (1979).

Today, with Sulzer diesel engines made only by licensees, their construction design and process continue to evolve in small steps. One thing that has changed significantly, however, is that many licensees subcontract a large number of their component parts, building about 30 per cent of each engine in-house. So, it is one of the marvels of the modern world that so many bits and pieces, manufactured in so many different places, can be assembled by so many diverse people, into so many complex machines, all of which perform the same.

Lasers are employed at the NKK shipyard for critical alignments.

In early 1998, RTA engines are delivering well over 25 million brake horsepower around the globe. We will now try to describe how another 74,700 horses were added to that incredible "stable." To do so, we'll take a bird's-eye look at the erection, in early 1998, of a 10-cylinder RTA96C, built by Hanjung (Korea Heavy Industries), in Changwon, Korea, for the 279-metre, 5,300-TEU, post-Panamax Hanjin containership *Hanjin Rome*. Although this description covers the process of constructing that engine at Hanjung in detail, the photographs that accompany it also illustrate the erection of the world's most powerful diesel, the 12-cylinder RTA96C built simultaneously at Diesel United in Aioi, Japan, as well as work being done at other licensees.

In fact, the basic approach to assembling a large diesel engine today is much the same at all licensees. They may use different methods and tools — the jigs and fixtures used in assembly are local choices, as are the lasers, theodolites, or spirit levels used to make the many precise measurements that assure the engine's alignment and optimal performance — but the results are the same.

The building of this great 1,740-tonne workhorse — which will truly function like a Swiss watch (if a bit more noisily) — was a culmination of 100 years of diesel know-how, insight, and caring. It began on 2 April 1997 when confirmation was received of Hanjin's order for two (plus five optional) 10RTA96C engines (the second for a sistership, *Hanjin Oslo*). The engines were ordered through Hanjung. As a licensee, Hanjung was already in possession of a full set of drawings and a bill of materials, so they immediately began ordering the components. Though subcontractors began making parts a year before the delivery of the engine, this engine would normally take about one and one-half to two months for actual assembly. But because it was the first RTA96C made in Korea, Hanjung logically moved with some added caution; the first engine eventually took about two and one-half months to complete.

The fabrication began with the welding up of steel plate for the two sections of the engine bedplate (one, for cylinders 1 through 4, weighed 82.2 tonnes; the second, for the

Among the first steps in construction of a large diesel is casting of cylinder liners, here shown at Bremer Vulkan's fiery foundry.

Hanjung's fabricating team makes a spectacular blue fire of its own as it welds up half of the bedplate for the 10-cylinder RTA96C engine.

The bearing shells are prepared *(top)* for laying-in of the crankshaft.

Once the crankshaft sections are set in and joined, the bedplate is ready for the column.

A true bird's-eye view: The bedplate is ground smooth and cleaned, ready for the completed column, which waits behind it.

remaining cylinders 5 through 10, weighed 94.4 tonnes). The two completed sections were joined and mounted on rails on the concrete floor of the main assembly hall. Beneath, a set of adjustable chocks firmly lifted and aligned the bedplate into an absolutely horizontal, or "zero," position, the basis for future assembly and measurement. The tolerance for the critical levelling of this massive structure is 0.04 mm per metre of engine length, to a maximum of 0.2 mm over the entire engine length.

The crankshaft came in two sections, corresponding to the bedplate sections: one weighed 132 tonnes, the other 172 tonnes. Before they were set in, the white-metal bearing shells were levelled in their respective girders. The crankshaft, with the fly-wheel, counterweights, and gear wheel (which will drive the camshaft through an intermediate gear), were lowered onto the bearings, and the crankshaft sections were bolted together at their mating flanges. Next, the bearing caps were bolted on, and the thrust bearing was installed at the engine's "driving" end.

Now, because more than 300 tons of crankshaft had been set into the bedplate, the "zero" alignment was readjusted, and

The bedplate level is reset once the weighty crankshaft is in place.

Crankshaft deflection is checked after bedplate levelling.

the crankshaft checked for deflection. At this point, Armin Probst, of field service, and Roberto Bianchin, from design (the Sulzer diesel "men on the job" who *stayed* on the job for the entire engine construction) were joined by representatives of the shipowner, Hanjin, and the classification societies, Korean Register and Lloyd's Register, who signed the protocol for the bedplate levelling and crankshaft deflection, required for classification. (Korean-owned ships often have two classification societies to assure the owners that they can resell their ships to non-Korean owners.)

Soon, the two column sections (60 and 85 tonnes respectively) were set in place atop the bedplate. They were micrometrically aligned to the centre of the engine. This was the first opportune moment for the outfitting crew to mount many external engine components, including platforms and galleries, electric cables, and piping. Next, the piston connecting rods were brought into the main hall, and each was lowered into the column (a delicate manoeuvre, as their shoes have a clearance as little as 0.2 mm to their rails). Each rod was pinned to its crank. All the while, outfitters added more

Probst and Bianchin in a rare moment of inactivity.

The column is ready for placement on the bedplate.

Once the column is cajoled into place *(top)*, the engine is made ready for the connecting rods, and the cylinder blocks are assembled *(above)*.

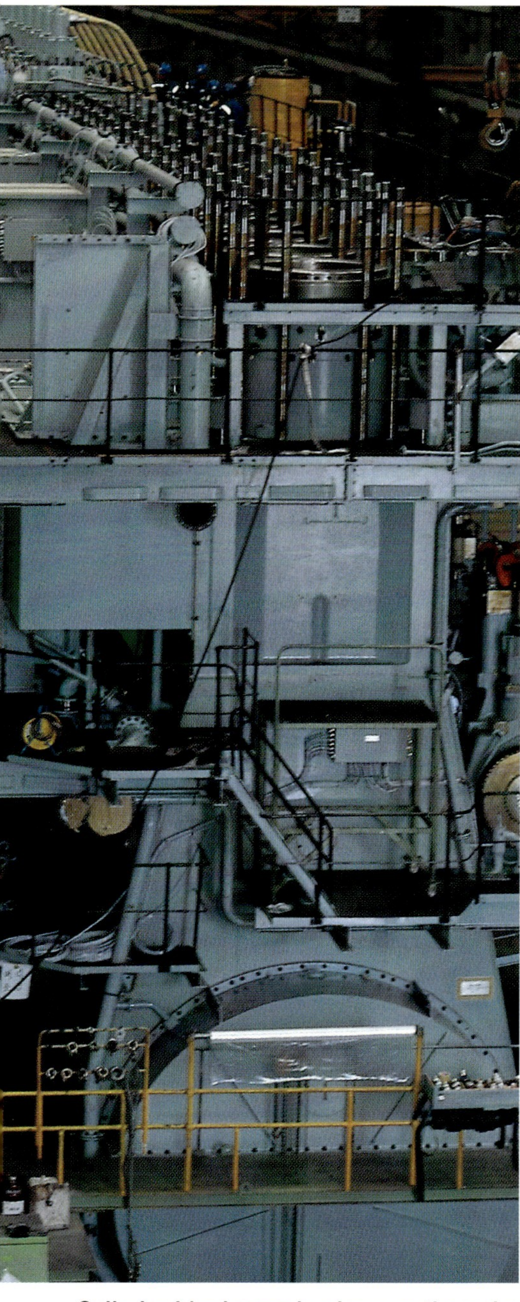

pipes, crankcase and column doors, and wiring.

Next, the cylinder jackets, with cylinder liners and water guide rings, were progressively bolted together as a unit, and precisely aligned to their own "zero" position. The cylinder covers were also made ready for bolting on (with eight bolts per cover). And, whatever gear the crew could conveniently mount at this time was installed on the jacket assembly, so the great crane could soon lift the entire works into place on the column. The complete scavenge air receiver was now mounted on the jacket assembly.

Meanwhile the cylinder covers were assembled, each with its three fuel injectors, starting air valve, safety valve, water guide ring, exhaust valve casing and exhaust valve, and connections. The covers were then temporarily bolted for a pressure test required for classification. The jackets were filled with water up the cylinder covers at a pressure of seven or eight bar, and left overnight under pressure. The next morning the entire assembly was checked for pressure loss, and leaks. The cylinders were drained and the covers were removed to allow for the three turbochargers (each weighing 12 tonnes) to be aligned and installed on the scavenge air inlet receiver.

Once it was in place, positioning pins were set to mark the turbochargers' precise alignment, so if removed later for servicing they can be re-fitted exactly into place. (In fact, all sub-assemblies that may someday need to be removed for servicing are located by positioning pins.) The turbochargers were removed to make it easier for the crane to transport the jacket assembly, which was lifted onto the column, and cautiously aligned. Again, the alignment of the entire cylinder block was checked, positioning holes were drilled and reamed for locator pins on the forward, centre and aft sections of the block, and the pins were installed.

The hollow pistons, each of which has dozens of blind holes drilled into it to receive "jet and shaker" cooling-oil nozzles, was

Cylinder blocks are in place on the column. Now the piston crowns are ready to receive the oil jets to complete the internal piston cooling systems.

One of the massive pistons, with its rod and stuffing box, is ready for insertion.

also pressure tested for classification at seven or eight bar, to assure that the "O" ring sealing the face between each piston crown (top) and skirt (body) cannot leak cooling oil into the combustion chamber.

Once approved, the pistons were set in their respective cylinders, an intricate process requiring a conical guide ring set atop the cylinder to firmly squeeze the rings into their grooves. Next, the crosshead and connecting rod joints were bolted together.

Now, the 9.5-metre tie rods, 12 per side, were inserted through the engine. Each was systematically tensioned in three stages — starting from the engine centre and working out — by a hydraulic jack threaded to its top. At first, each rod was tensioned to 100-bar pressure, its nut hand tightened, and its elongation measured as a check against tension. The second stage brought the tensioning up by 600 bar, the final by 1,000 bar. The nuts were tightened, and the final calculation of elongation was made to assure that it met the standard of approximately 11.5 mm per rod. Once the entire cylinder cover assembly was set in place and hydraulically bolted down, a crew of coveralled men entered the crankcase to tighten down and lock the bolts for the main bearings, connecting rod bearings, and piston. Then the piping crew swarmed over the engine to install the remaining

Machined tie rods await setting into the engine.

Hydraulic and electrical tools are used for drilling and reaming holes.

pipes for cooling water, fuel, and pneumatic system controls.

The crankshaft, intermediate, and camshaft gears, which impart rotation to the camshaft, were adjusted in sequence, to be parallel to within 0.03 mm, with clearance (backlash) between 0.4 mm and 0.78 mm. Next, the five fuel pump and exhaust valve actuator blocks (one unit for each two cylinders) were aligned to their drive gears within 0.02 mm, and the piping crew completed fuel line installations. The three huge turbochargers were then re-installed, and the exhaust manifold was mounted and aligned to the cylinder covers.

In preparation for the bench test and running-in,

An inspector prepares to check the assembly of the gear drive.

a subcontractor connected up cooling water, fuel, and lubricants to the engine, and the exhaust manifold was hooked up to Hanjung's external exhaust system. Meanwhile the erection crew checked that each piston was centred in its liner and that the crosshead guide shoes ran evenly on their rails. All the parameters of alignment were re-measured in three randomly selected cylinders, and the the next phase of the protocol was signed. Then, with all engine clearances and alignments found to be within tolerance, the completed protocol was signed by the owner Hanjin Shipping, Hanjung, Wärtsilä NSD, and the classification societies. Now Hanjung was responsible for the engine until the completion of the shop tests and the signing of the final protocol. The engine was ready to be fired up.

To get absolutely ready for that magnificent moment, the Hanjung engineers cleaned the engine, flushing it with oil to remove welding deposits, dirt and other possible residues. This extremely important process can take two or three days, as no foreign particles can be permitted to remain within the engine or its pipings to enter the bearings and seals and risk damage. Then, a water dynamometer was aligned to the flywheel end of the engine, as a substitute load for the propeller.

Oil is drained from a piston rod after its pressure test.

The engine is prepared for test and running-in.

The world's most powerful diesel engine, the 12RTA96C, ready for shop testing at Diesel United in Japan.

Notes are taken, data are recorded, during the running-in at Hanjung.

Along with many other items on the long shop test checklist, the fuel pump timing was adjusted against the crankshaft angle. Control air was supplied to the piping system to check the control system and valves. And high-pressure starting air was forced into the cylinders to turn the engine over for a moment, to clean the cylinders and blow out residual oil lying in the concave piston crowns.

Finally, after more than two months in which this RTA96C grew from inert, unbending, welded metal, into a glowing tower of potential energy, it was ready to have its preliminary shop test. It was ready to become an engine.

The first running of a new engine is always witnessed privately by Wärtsilä NSD and the licensee, and also follows a strict protocol. The 30-bar starting air is injected once more into the cylinders. The engine turns over and after a few revolutions fuel is injected. Suddenly, there is fire and rumbling in the belly of the beast, and at last the great machine begins to run under its own power.

It is a truly miraculous moment, all those parts, from so many places, fitting together — and with a roar the engine works!

The test protocol calls for the engine to be run for eight to 10 hours, while its speed and load are slowly built up. First, it is run at 15 per cent contracted maximum continuous rating (CMCR) for 20 minutes, then it is shut down for engineers to enter the crankcase to check crankshaft, connecting rod and crosshead bearings for overheating by touching each by hand. They check the gear wheels and turbochargers; they look for leaks in pipes.

The team closes the engine and restarts it, running it for one hour at 25 per cent CMCR load. The engine is again shut down and gear wheels, cam surfaces, and scavenge air manifold are checked, and experts peer through the scavenge-air ports looking for scuffing or piston ring damage. They close the engine once more, and progressively repeat the cycle — running, shutdown, and inspection — through 50, 75, and 100 per cent CMCR, until they run the engine at 110 per cent CMCR load for a short time. During runs, engine parameters are recorded and continually analysed; during shutdowns, cams are checked, turbochargers assessed, fuel injectors inspected.

At the final shop test, the owner, classification societies,

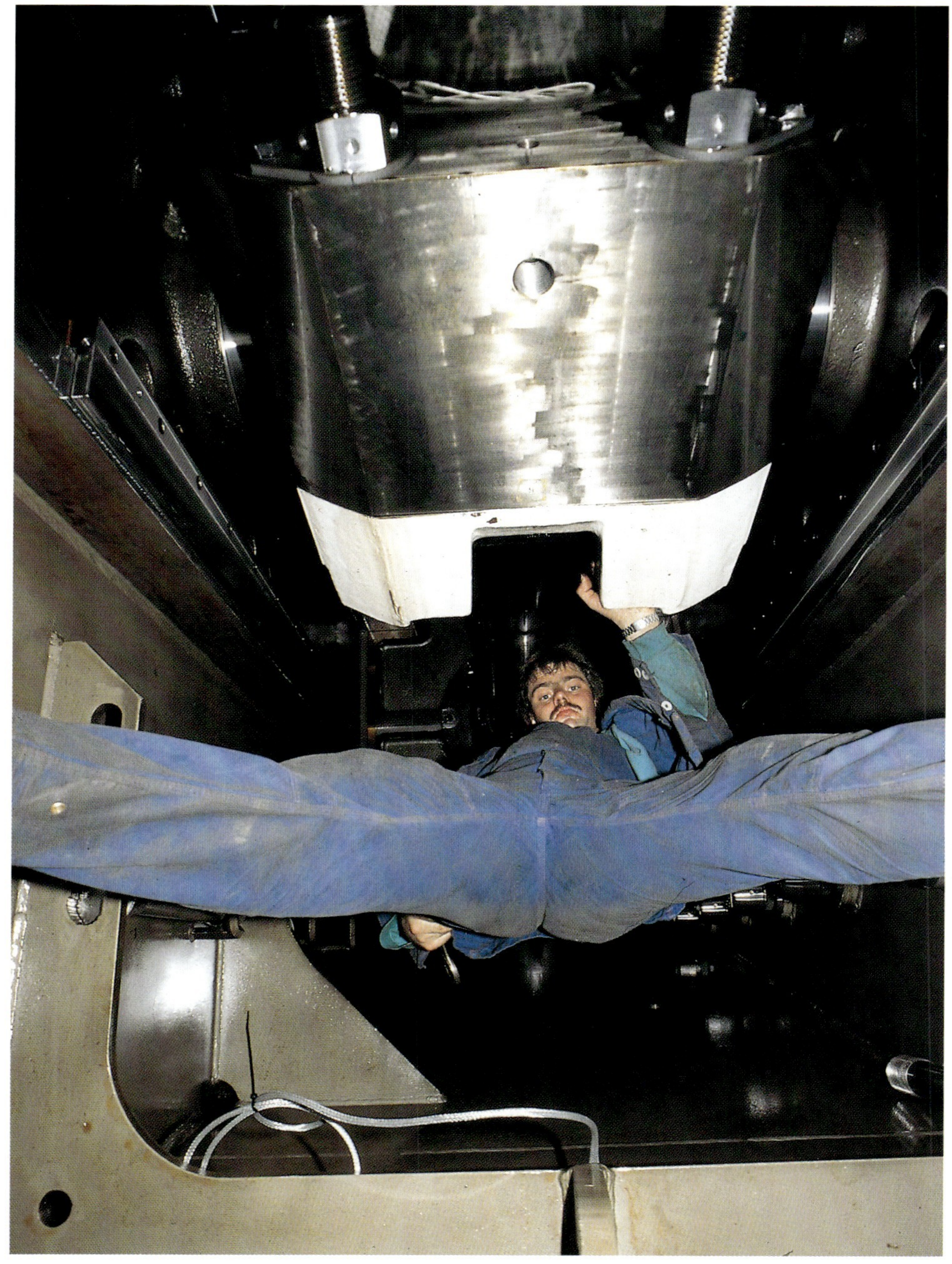

Repeated inspections during testing and running-in are mandatory to assure one Sulzer diesel quality no matter which licensee manufactures an engine.

Once shop tests are completed, the engine is transported and installed in its ship, as here at Ishikawajima-Harima Heavy Industries.

Wärtsilä NSD, and Hanjung are present. This is a day-long test, repeating the running-in test to 110 per cent. Tests are run on all alarms (temperature, cooling water, oil pressure), as well as on the governor; an emergency stop is run. If these test pass all requirements of the protocol, it is signed by all parties and the classification societies give the engine a stamp of approval on its nameplate. The engine is now accepted for installation on the ship.

Final measurement is made on crankshaft deflection, and the protocol is handed over to the owner, while Wärtsilä NSD Switzerland takes all test results and clearance tables for the records it keeps of the history of every engine it supplies.

Once the 10RTA96C built at Hanjung was approved, the engine was dismantled into major components, packed in protective material, placed on barges and taken upcoast to the Hanjin Shipyard in Pusan. There the engine was reassembled in the ship under the guidance of a Wärtsilä NSD representative, who also attended the sea trials of the ship and sailed on her maiden voyage.

As you complete the reading of this essentially brief essay on engine building, somewhere out on a dark, perhaps stormy ocean, *Hanjin Rome* and her sister *Hanjin Oslo*, along with their bigger, more powerful sisters from P&O Nedlloyd and NYK, are nobly plying their trade. And deep beneath their decks, loaded high with containers filled with worldly goods, a Sulzer RTA96C "Swiss watch" — albeit a noisy one — is turning a great propeller, spiralling at 100 rpm, to assure that these ships keep their rendezvous with bustling container ports to the East and West.

Applications

When a Sulzer diesel licensee hands an engine over to a shipyard it has full responsibility for the engine's quality, but only so far as the naked engine is concerned. The licensee bears no responsibility for proper installation of the engine, which is in the hands of the shipyard. Nor does it assume liability for peripheral systems that supply the engine with life-giving fluids — fuel, oil, water, air — and electricity. That is how it works in theory.

But, in real life, if trouble develops during the shipyard's guarantee period, and even after the one-year guarantee runs out, the shipowner always calls Winterthur. So, Winterthur engineers do not simply walk away from a project when the

Engine peripherals are installed before the main engine is set in its ship.

The shipyard is responsible for the installation, piping, wiring and more.

licensee has delivered an engine to a shipyard. Wärtsilä NSD has a very high investment in each engine, and has compelling reasons to see that its installation is completed without a snag, as the company's reputation is on the line regardless of who is legally liable for an engine. And it has equally compelling motivation to see that peripheral systems are chosen and installed optimally as well. This gives greater assurance to the shipowner that, if properly maintained, his engine will work well, and that his ship's resale value will remain reasonable.

The Application Engineering department, under the gentle but firm guidance of Frédéric Porchet, therefore, acts as liaison between licensees and shipyards, to assure trouble-free installation and correct selection of peripheral system components. The documentation people travel often to shipyards — acting as consultants, diplomats, occasionally referees. They provide shipyards with drawings, specifications, and tolerances to assure proper seating and alignment of engines, application of starting air, cooling water, lubrication oil, and fuel supply, and work closely with shipyards in aligning propeller shafts, propellers, and bearings. Early in a ship-design project, Winterthur engineers also give their best recommendations on the shape and content of an engine room, acting as an additional pair of eyes to see that the ancillaries are set compatibly with the ship's hull form and structure. And they recommend the proper type of auxiliary engines (preferably Wärtsilä or Sulzer auxiliary engines) to supply the ship with electrical power. In short, the application and documentation people work hard to help reduce a shipyard's engineering effort. In addition, the department works with classification societies, to understand their viewpoints. (For example, the American Bureau of Shipping, Lloyd's Register, and Germanischer Lloyd leave some room for interpretation of their standards, but the Russian classification society, with its history of strict military applications and little commercial experience, leaves no room.)

Technical services

A shipowner would like to be married for 20 to 25 years with each ship in his fleet (unless it is sold or sunk). If one of his ships is routinely sailing in mid-ocean, however, and the chief engineer suddenly wonders if the engine is beginning to sound funny, where does he call? Winterthur, of course; at any time, 24 hours a day there will always be a comforting voice "at home" on the Sulzer diesel hotline. Winterthur supports its clients with manpower, because shipowners are most comfortable when they know that, in case of trouble, they can get "the Sulzer man" on the job.

There are 70 service-oriented "Sulzer men" in Winterthur, more than 50 of whom are on the road most of the time, troubleshooting ship problems, checking ship maintenance, following up on guarantee work, and getting ships back on the line. The service staff keeps a record of every engine built, whether in Winterthur or by licensees, noting the ship (or power plant) in which it was installed, its maintenance and overhaul history, breakdowns (even its demise with the scrapping of the ship). Early records were hand-written on file cards, later typewritten. In 1983 Sulzer developed an appropriate electronic database, which is now kept filled through regular questionnaires to ship engineers. Using the data, the Winterthur service men are in a position to make recommendations to the ship for proper maintenance. If the owner sells the ship, these records can help the buyer determine the ship's condition.

But global service-call work is the most important function of the technical service department. Most service calls concern routine matters but, as the following cases demonstrate, disaster can occasionaly loom in a ship's life, and it is at those times that a field service engineer really earns his keep.

A typical case took place in December 1996. A 3,000-TEU containership with a 7RTA84 limped into Damietta, Egypt, north-west of Port Said. Her engine was frozen: liners were split; pistons cracked; rings broken or melted. Hardly any cooling water or lubricating oil was reaching the engine, which had severely overheated. The cause: the gensets were running well below speed, thus the oil and water pumps were running at half load. Although a number of alarms had gone off, the engineers ignored them, allowing the engine to run merrily along at "Full Ahead". The Sulzer field man arrived, assessed the situation, called Winterthur, and ordered 40 tonnes of parts —

"I just have time to catch the next flight!" (Sometimes, a field service man's life is no "picnic".)

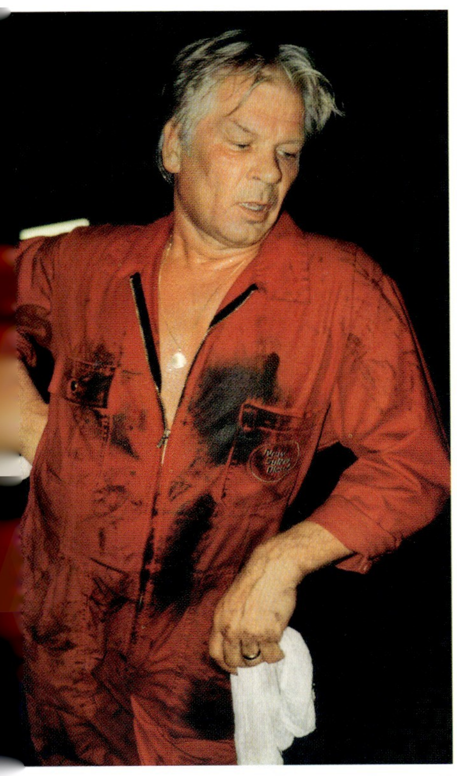

The "King of Service," Klaus-Dieter Muuss *(left)* and his Winterthur-based colleagues are not afraid of a little heat and grease, so long as they keep those engines running.

reconditioned liners, pistons and rings, and other spares — from several sources, mostly Rotterdam, which were flown in by charter aircraft. He had them expedited through customs (with some *baksheesh* to smooth the way) and saw to their installation. Within a week the ship resumed its voyage. When another box ship broke down in Hawaii, with severe scoring in its cylinder liners, service engineers did a jury repair by dressing the liners with grinding stones to temporarily reduce blow-by, so the ship could proceed at low power to Kaohsiung, where replacement parts were installed. And service engineers were called to a cruise ship that had been refit with a new Sulzer engine, but still had its old fuel system (designed for a steam turbine). Ultimately that ship had to be taken out of service for two weeks to replace the system, at a loss of US$ 200,000 per day, but Sulzer engineers solved the problem.

Klaus-Dieter Muuss (who saved all the ships above) is Winterthur's longest-serving field engineer (friends and shipowners call him "The King of Service"). In 1981 Muuss was summoned to Tahiti by the famous cruise liner *Sagafjord*, on a luxury world cruise. With no warning (ring- and liner-wear sensors were broken), blow-by from worn piston rings had reduced power in one of her two main engines. As the ship kept to her cruise schedule, Muuss reached into the engine and estimated the blow-by by hand (again, there were no sensors), then had the engine shut down for further inspection. On the way to Rarotonga, he determined that incompatible bunkers and a missing fuel filter had precipitated the breakdown. Installing

Old liners never die, they just become planters *(left)*. Transportation to the ship is by any method you can find. A Sulzer brand engine is more than a machine *(right)*.

an old fuel purifier, he filled the day tank with clean fuel. In Sydney, preliminary repairs were done, then a fresh crew followed the ship to Hong Kong where, over four days, parts were overhauled by skilled Chinese artisans with hand tools working in the streets. In Singapore and Bombay, "Sulzer men" continued repairs and by the time *Sagafjord* steamed into Piraeus, for her guests to go dancing in *Plaka*, her engine was normal. A rather unusual service call, to say the least.

Something else is unusual about today's service calls. In the old days the service man carried a basic tool box, some clean brown coveralls and some blueprint drawings. Now he needs a laptop to check injection and exhaust valve timing, a microscope to examine liners, and a magnifying camera to take super close-ups of "cat fines" (abrasion fuel residue from the catalytic cracking process). And he needs a modem hooked up to a telephone line to gain access to Winterthur's engine database and to read his e-mail.

With all the pressure in Winterthur for new and better products, it is important for field personnel who perform these remote tasks to bring their experience to the home office. So, once each winter, just after the New Year, service engineers meet in Winterthur to exchange know-how, hold seminars on their field experience, and break bread together. This business is not just about tribology and electronics; it is about relationships, inside the company as well as outside. As Leo Schnellmann, head of the Service Department, proudly proclaims: "Behind the spanners, the drawing boards, the laptops, and the grease, there are *real* people."

Spares

The spares department services an incredible number of ships with Sulzer engines — about 6,000 with two-stroke, about 8,000 with four-stroke — plus countless stationary power plants. There are more than 50,000 part numbers on a typical two-stroke marine engine, with many times that number in total parts, and no ship can ever expect to carry anything approaching a full inventory of spares. Wärtsilä NSD is prepared to supply them all — fuel injectors, piston rings, stuffing boxes, nuts and bolts — through its worldwide sales network and its primary supply warehouses in Winterthur, Singapore, Rotterdam, and Miami.

This network is maintained by the head of the spares department, Werner Jungblut, who sees to it that Wärtsilä NSD Switzerland supplies spares directly to shipowners for all of its engines, even the very old ones. Licensees sell spares also, but only for the engines they build. And many chandlers around the world sell spares. The spares business is a highly competitive one. It is also a highly lucrative one, but with many potential headaches relating to interchangeability. (Through all its incarnations, for example, the RTA58 has undergone many injector modifications; without perfect record keeping it is difficult to know which model to replace on any given engine.) Occasionally, licensees make small modifications in parts, as a selling point. If the changes are purely cosmetic, and don't affect interchangeability, that is no problem. But for operating parts they must get authorisation from Winterthur, or there would be chaos in the field.

Usually, replacements are sold to shipowners after the normal wear and tear of the old parts. But if a part has to be re-engineered because of a design fault, it is supplied at no charge. An example of this was the piston rod stuffing box on some high-horsepower RTA engines. This gland box, a handful of bronze or Teflon rings, separates the piston underside ("dirty" space) from the crankcase ("clean" space) by scraping oil from the rod and distributing it to various tanks. The stuffing boxes began to wear the rods rather too quickly, resulting in poor seals, loss of lubricating oil, and contamination of crankcase oil. This problem eventually developed on 500 vessels, which had to be retrofit with a modification package.

Shipowners buy only 25 per cent of their spares from Wärtsilä NSD Switzerland; they buy the rest from other suppliers who compete with lower prices because they do no research, no development, no training. And some parts on the market are pirated (copied) by unauthorised parties and sold at much lower market prices. Strict control over part specification is a way of fighting pirated spares.

An engine can run only on a ready supply of fuel and spares.

Today's trends

Approximately ten to fifteen per cent of a ship's construction cost is in its engine. All shipyards are looking for economies of scale and construction standardisation to keep ship prices from escalating. Over the coming years there will be an increase in normalisation of engine room layouts, and more modular ship design and construction, for shipyards to apply more mass production methods. The resulting decrease in ship cost per tonne will help the transportation industry, which is under tremendous cost-cutting pressure from product shippers. Wärtsilä NSD is therefore dedicated to helping its licensees reduce engine costs to hold the line on the first cost of new ships.

The strategic thrust of Wärtsilä NSD Switzerland's new marketing programme for the Sulzer brand is to increase its market share in tailor-made engines for standard ships, while maintaining its traditional strength in the market for large containerships. One way in which the creative people of Winterthur have already accomplished this is in the successful completion of the programme bearing the intriguing names of "The Pegasus Project" with its sub-project "Shoehorn." The Pegasus Project, originating in the mid-1990s, was aimed broadly at bringing new Sulzer diesel engines to the market, rather than the old style of making the market come to Sulzer engines. After lengthy consultations with major Japanese and Korean shipyards, New Sulzer Diesel developed important features the yards sought in tailored engines for standard ships, which resulted in the optimisation of the RTA48T and RTA58T for the standard type of bulkers and tankers in demand.

Shoehorn, spearheaded by Jürgen Gerdes, a ship engineer by training, now manager of Marketing Development, was conceived to help shipyards reduce engine room size, to lower costs and provide greater volume for income-producing cargo, by methods such as reducing engine dimensions, moving engines aft in the hull, modifying the transmission of propeller thrust, improving the engines' ancillary systems and much more.

At the end of the 20th century, at the dawn of a new millennium, the Sulzer-brand diesel engine is "setting its sails" before the winds of tomorrow, and delivering a message of strength and confidence in the future.

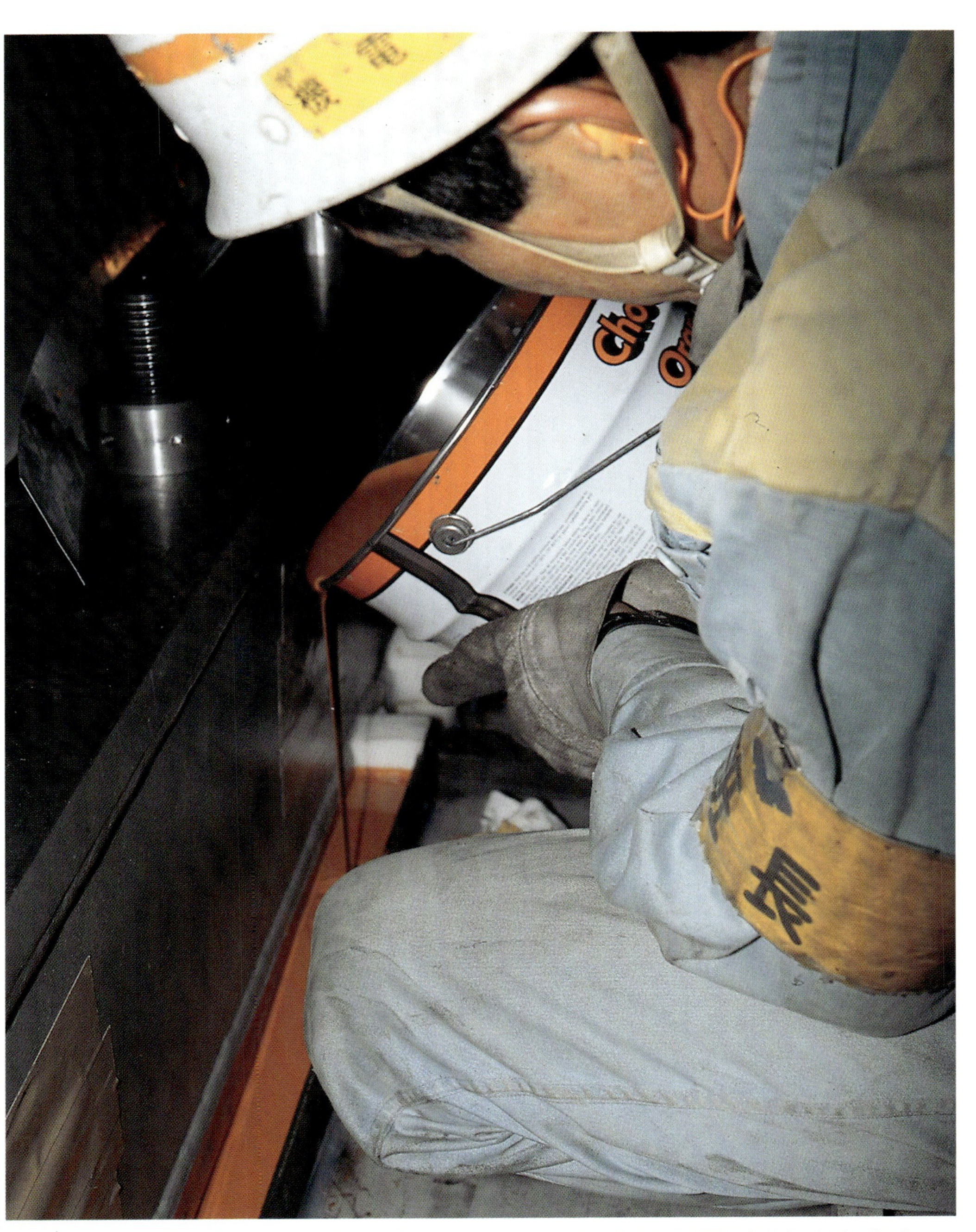

"Shoehorn" initiated epoxy-fitted thrust sleeves to distribute engine loads to the hull.

Star Sea Bird, the first vessel to have an engine ideally optimised through the Pegasus programme: a six-cylinder RTA48T Sulzer diesel.

Tomorrow

By November 1990, when Sulzer Brothers sold the bulk of its shares in the diesel division, the Sulzer brand was well established as a prime contributor to the advancement of diesel technology. Once the ferment had been quelled, in fact, the early 1990s turned into surprisingly golden ones for New Sulzer Diesel. There was relative health in marine engine sales, particularly with the RTA series engines announced in 1982, which were well on their way to becoming standard propulsion for highly sophisticated vessels. The early success of the RTA84C, the company's "queen of the fleet," boded well for the future of the firm's ability to continue to supply large engines for containerships. But in particular, New Sulzer Diesel's power plant division had an especially fine period: for some two or three years after the diesel division sale it delivered more than 50 turnkey power plants to China for it's burgeoning industrial base and growing civil population. (These complete systems were manufactured by New Sulzer Diesel's French daughter company, near Paris, a specialist in power plants. The transaction was facilitated by a consortium of French banks that had the patience and perseverance to work with the Chinese authorities.) Applying a portion of the profits from those sales, New Sulzer Diesel was able to take a major step in the establishment of its new image and realisation of a long-held dream.

For many years Sulzer Brothers tested engines in rather basic facilities in Winterthur. But despite the age of its buildings and the limits of its facilities, the company was able to test some advanced engines, including the RTX "Technology Demonstrator," the world's first electronically controlled, hydraulically operated two-stroke crosshead engine (which didn't use its camshaft). That engine, which ran for more than 2,500 test hours in Winterthur, before the old facilities were closed, led to new frontiers in diesel technology, and gave life and breath to many developments that now are integral parts of the Sulzer diesel line.

But after Sulzer ceased making engines in Winterthur in late 1988, it became abundantly clear to the design and research teams that the company sorely needed modern facilities to develop and test new engines and components, if New Sulzer Diesel's research was to be less dependent upon licensees' test

The RTX "Technology Demonstrator" was dismantled, and moved to the Training Center.

beds, which are necessarily geared to production. In 1993 the New Sulzer Diesel board decided to fulfil that need by building a research facility close to its home offices in Winterthur, which became the Diesel Technology Center.

Ground was broken in October 1993, by Dr. George Lustgarten, then Senior Vice President, at the very time that New Sulzer Diesel was about to make a great stride forward in engine power with the creation of the RTA96C. The Diesel Technology Center building was completed only one year later, in October 1994, just two months before the Sulzer RTA96C was announced to the shipping world. The Center, which subsumes some 40,000 cubic metres of volume and covers an area of about 3,500 square metres, is located in the suburb of Oberwinterthur, in an industrial area that is largely occupied by manufacturing divisions of Sulzer Ltd., New

Ground was broken for the new Diesel Technolgy Center in October 1993, and the first earth-shaking fanfare was heard at the Center just two years later.

Sulzer Diesel's former "parent". After the completion of the structure, engineers required another year to install the test beds, control rooms, and machine shop needed to turn this empty building into a working research and development facility. But when it was completed, it obviously stood as the most advanced diesel research facility in the world.

The Diesel Technology Center was opened to great fanfare in October 1995. Over several days of celebration, bankers, industrialist, politicians, shipowners, licensees, shipyards, the press, friends from Switzerland and afar, and the company's executives, employees, and their families, congregated in the main hall, which had been converted into a combined theatre and dining room to mark the occasion. It was a wonderful international gathering of people who came to share in this magnificent achievement, and to join in

This delicious five-cylinder "Confectionary Demonstrator" was the first engine to "run" at the Center, accompanied by melodies played by the Sulzer big band.

The celebratory candlelight dinner was a perfect blend of the joy of delicate cuisine and the promise of heavy technology.

expressing confidence in the future of the Sulzer brand. As the Sulzer spirit had long ago ordained, this person-to-person contact, this human side of diesel engineering, especially at the time of such a forward-looking celebration, brought an added degree of warmth into a business that can otherwise seem basically to consist of little more than nuts, bolts, clamour, and competition.

After the party concluded, once the dining room and theatre were dismantled, and guests all returned to their homes and homelands, the serious work of the Diesel Technology Center commenced. Let us look at the facility and its functions in its first few years of existence. Let us not, however, look first at what the Diesel Technology Center *does*; let us rather start by looking at what it *does not do*: It *does not* pollute the sweet airs of Switzerland. As anyone who has driven on the motorway behind a diesel truck in low gear on an upgrade can attest, diesel engines can be capable of producing emissions of all sorts: noxious oxides of nitrogen and sulphur, smoke, and soot. In order for the researchers to enjoy unfettered experimentation on all sorts of diesel technology — including unproved technology that is certain to produce worse than normal pollution — and to test emissions-reduction systems themselves, it must first be assured that none of the polluting exhaust gases and particulates from test engines can enter the atmosphere and impinge on the environment. So New Sulzer Diesel installed a complete, and quite expensive, after-treatment plant, with advanced catalytic converters, scrubbers, and filters for each test bed. This system, wholeheartedly approved by the local government, prevents any and all engine exhaust pollutants from passing into the outside air. Installing this great system, of course, was not alone a moral decision; it was based on strict Swiss anti-pollution laws, which result in one of the cleanest countries on Earth. The towering metallic exhaust system has virtually become the trademark of the facility, as it can be seen in the clear Swiss air for quite a distance, against the distant backdrop of the snow-covered Alps, in the otherwise flat valley where the facility stands.

Within the building, since its 1995 inception, work has continued in a number of areas of development. Primarily, it is a test bed for large two-stroke,

The Center's sculptural signature: the towers that remove all engine exhaust pollutants.

low-speed crosshead engines, and large medium-speed four-stroke engines. Because of the exhaust treatment that each engine receives, it can be run routinely on marine diesel fuel or, more importantly, on heavy fuel oil of the poorest quality, without restriction, so that engines are tested under real-life conditions.

This is a very important aspect of diesel engine design research, because of the enormous variation in bunker quality throughout the marine world — a good deal of it being nothing short of dreadful.

Each of the test beds at the Diesel Technology Center is adjoined by an enclosed control room equipped with advanced monitoring and measurement equipment, selected and arranged for researchers to quickly gather and analyse test data for use in the design of new engines and modification of old ones. A separate laboratory is dedicated to testing smaller engine components and systems.

The Center's complete machine shop has become an ideal source of manufacture of test components, as well as an equally valuable location for developing component manufacturing processes to be shared with licensees.

The first major accomplishment of the Diesel Technology Center was the validation of the design of the Sulzer RTA58T, a low-speed engine with 580-mm bore and 2,416-mm stroke, rated at 2,720 horsepower per cylinder at 103 rpm. A four-cylinder version of the engine, manufactured by the German licensee Dieselmotorenwerk Vulkan AG, in Rostock, was installed in the test facility before its opening. A thorough testing programme of that engine resulted in the confirmation of all its thermodynamic parameters, and paved the way for further successful tests, and subsequent production, of both the RTA58T and RTA48T, both tailor-made Sulzer engines that were designed specifically for service in bulk carriers and tankers with cargo capacities between 20,000 and 90,000 deadweight tonnes.

The RTA58T benefitted greatly from research work done with the RTX Technology Demonstrator, and was at the time considered by Sulzer to be an "interpolation" engine, a sort of cross between the existing technology of the RTA series engines and the RTX, which for experimental purposes was run at much higher loads and pressures than today's engines are asked to do.

A four-cylinder RTA58T, the first engine to be validated at the Diesel Technology Center.

The eight-cylinder ZA40S was also transferred to the Diesel Technology Center.

Another engine, brought over from the old Winterthur test facilities, installed and used extensively in the new Diesel Technology Center from the outset, was an 8ZA40S. This medium-speed, four-stroke engine, which was introduced in 1985, has a 400-mm bore, 560-mm stroke, and develops 7,840 horsepower at 510 rpm. The ZA40S became one of the most successful engines of its type in the market, and was the leading four-stroke engine in the Sulzer diesel catalogue until the development of the ZA50S, whose six-cylinder version has taken its own place in history.

Both the two-stroke and four-stroke development work performed at the Diesel Technology Center are geared toward enhancing engine reliability and fuel efficiency, and lowering manufacturing costs. Along those lines, researchers perform extensive parallel testing of engine components and subsystems, such as fuel injection systems and exhaust valve drives, to assure their functionality before they are applied to test or production engines, as a way of speeding development.

The other important side of the Diesel Technology Center is its well developed training programme. A group of highly qualified instructors, headed by Paul Krebser, all of whom are senior engine servicemen themselves, train engine operation engineers in the ways, means, and riddles of diesel engine technology. Visiting engineers get an introductory course on the engine they will service, with emphasis on how the engine works, its maintenance requirements, proper component operation — the whole gamut of diesel engine knowledge.

When a shipyard signs a contract to build a group of ships for a client who is new to the Sulzer diesel brand, the Training Center welcomes the shipping company's engineers and also trains them for one or two weeks, a service that is provided to the shipowner, including board and lodging in Winterthur (and occasionally a token gift of pocket money for those on a limited budget).

The Center also assists in on-the-job training for ship engineers, by sending qualified instructors into the field to give hands-on instruction aboard ship. Through the Center, and the service department, Wärtsilä NSD sends qualified engineers to accompany a ship on its maiden voyage and beyond, not only to give the engineers final instruction and

The Technology Center's prime activities include mastering the diesel engine and training those who construct and use it.

added assurance, but to observe firsthand the long-term operation of their engines. And some friendly shipowners even permit Winterthur engineers to install new components aboard their ships to observe their operation in the field.

Important training is also offered to Sulzer diesel licensees, who send engineers and production people to improve their skills. For example, Wärtsilä NSD has taught key personnel from the Chinese licensees, who were relatively inexperienced, in the fine arts of metal cutting and welding, skills which they brought back to their plants and shared with their colleagues on the job. (In one typical case, a group of welders from Dalian Shipyard, in China, came for training in Winterthur to improve their skills in the highly sensitive process of bedplate and column fabrication. When the welders returned home, indeed the quality of their work took a good leap forward. Quality is a matter of investing in higher education.)

In addition to its sophisticated equipment, methodology, and training, the Diesel Technology Center has a very important social benefit. When engineers from shipping companies or welders from licensees come to Oberwinterthur to train, they go through an intensive experience — sometimes an overwhelming experience. This must be relieved, occasionally, by no less intensive entertainment and tourism. Visitors having their first experience in Switzerland cannot be denied their request to view the Matterhorn or tour Geneva, or miss an opportunity to buy Swiss cuckoo clocks (which, by the way, are mostly made in the Black Forest of Germany).

But much has changed in recent years. The Chinese delegations, which once preferred to return home through Hong Kong to buy colour television sets, are now more interested in perfect welds; they are all business. So the Diesel Technology Center is much more than a test bed for engines and a training faculty.

It is, perhaps above all, an all-important motivating vehicle for technical people in Winterthur to stay in friendly contact with their market, their shipowners, and their licensees, to give them a helping hand in technology and manufacturing prowess — in the interest of license care, market care and, finally, shipowner care.

No great company can sell great engines on technology and price alone.

Diesel technology today

Though its prime activity is research, the Diesel Technology Center is not just a monument to science, engineering, and sociability. It is an active facility, under the direction of Iso Karrer, engaged in answering questions about the future of the diesel engine. And the future of the diesel engine is closely linked to the future of naval architecture and shipping around the world.

One of the prime consequences of ships carrying crude oil, liquid chemicals, dry ores, bulk materials, liquefied gases, refrigerated produce, and containerised goods (not to mention high-paying, well-dressed cruise passengers), is to drive naval architects toward new and better ship designs. (The prime legacy of the 1967 Suez Canal closing, for example, was to force architects to draw larger-capacity tankers, a few with more than 500,000-dwt capacity, which, as a result of the late-1990s oil glut, are all but impossible to conceive of today.)

Naval architects are in a constant search for more efficiently shaped ships that can be more easily driven by existing engines. Their goal is to progressively reduce hull resistance and loss of energy into

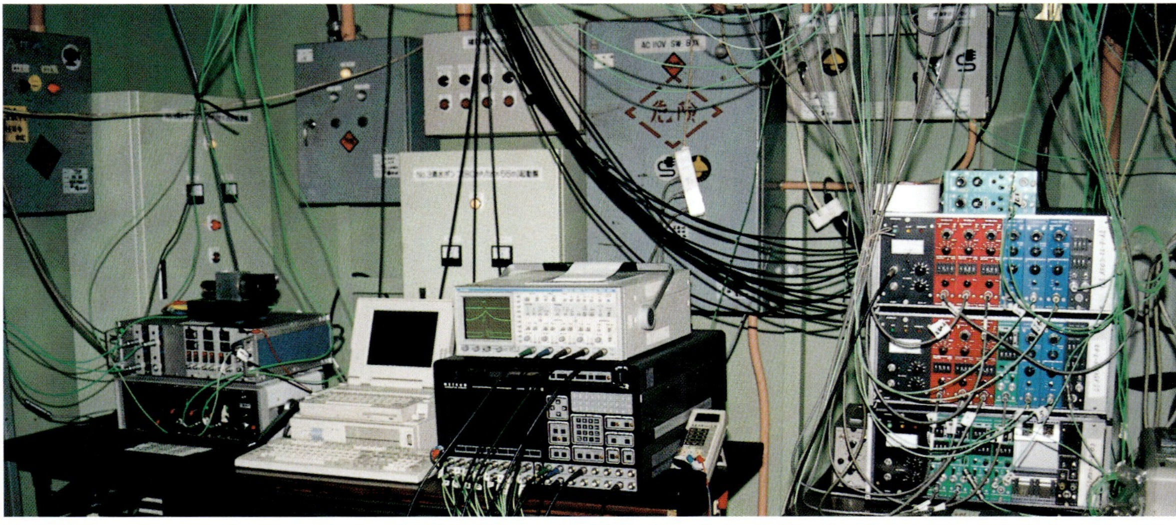

Not much has changed since 1936 *(top)* with respect to handy wiring for diesel research.

the ships' wake, so that the net efficiency of the drive train and the ship are incrementally raised. (With competitive, high-speed ferries, naval architects have accomplished some of the most fascinating results, with hydrofoils, catamarans, and wave-piercing hulls.)

In addition to tank testing models with modified hull forms, new bulbous bows, asymmetric sterns, and new rudders, ship designers of large commercial vessels are trying to squeeze the smallest fraction of a knot and the smallest percentage of efficiency out of their designs with improved propeller shapes, contra-rotating and variable-pitch propellers, propeller tunnels and nozzles, wake-enhancing bulbs, thrust-enhancing vane wheels, propulsion pods, even waterjets. Yet, with the reality of all this research, and the specialisation of these vessels, there seems to be nothing better than a good old fashioned, well-designed displacement hull, with a good clean bottom, driven by a good old fashioned diesel engine.

In recent years, now that the largest tankers have returned to the more reasonable size of 200,000 to 300,000 dwt, some sense has returned to that sector. But, tankers aside, there is an undeniable trend toward ever larger containerships. It would appear that, even with containerships already carrying more than 6,000 TEU, the limit has not yet been reached. (Indeed, nobody knows what the limit is. There are designs for 8,000-TEU single-screw containerships, which may have to wait only for larger, deeper and more mechanised container ports.)

It is much the same with diesel engines. Nobody knows what the limit is. A few years ago, many engineers could not imagine a marine diesel delivering more than 70,000 horsepower — it was no magic number, just a very challenging one. Now we have diesels that approach 100,000 horsepower, and there is no limit in sight. The (hypothetical) 8,000-TEU containership, with service speed of 25 or 26 knots, will require a 110,000-hp engine. (Then, we can no doubt expect containerships with greater than 10,000-TEU capacity coming over the horizon.)

Today, we know that one development required for shipowners to be willing to order these bigger ships, with their draught limitations in harbour, is a new generation of propellers designed to accommodate much more horsepower without an increase in their diameters — propellers that are capable of higher blade loads, deliver more thrust from each blade, and are made of newer,

Evergreen's *Ever Decent*, with a Sulzer 12RTA84C engine, is a prime example of the development of the containership type.

stronger materials. And it probably also means engines that run at higher revolutions.

If ships continue to get larger, especially containerships, designers must reconsider twin-screw propulsion, although shipowners are not pleased with the prospect of the added capital investment of two complete engine systems (there are already prospects for a 13,000-TEU twin-screw containership). So long as these ships are driven by single-screw propulsion, however, producing the very large engines that will drive them is certainly within the realm of possibility for the Technology Department. Their attitude is: Nothing is impossible. There is no problem developing diesel engines that exceed 100,000 horsepower. And there is good incentive to do so: a single engine with a single propeller is still the safest, cheapest way to drive a large ship.

Kaspar Aeberli is Director of Technology for Sulzer diesel engine development. He is well suited to his task, as he seems to embrace diesel engines with a rare passion, as though they were animate objects — not necessarily an easy thing to do. It is Aeberli's responsibility, among others, to see into the future, beyond current limits (real or imagined), to give the market what it needs. Aeberli is a thoughtful man, but he has no illusions about market realities to come. To accomplish his mission, the Technology group has a staff of some 160 people performing research, testing, design, and engineering. Six of those engineers are in a group called "pre-development," a sort of semi-ivory tower in which they are charged with gazing at the "big picture" and finding ways to make engines run better. That department is run by Peter Baumgartner, a long-time engineering advocate of four-stroke engines, and still their chief advocate within the company, who has willingly applied his experience to new two-stroke development.

Under the Technology umbrella, Research & Development, led by Nikola Mikulicic, does advance work in thermodynamics, fuel injection, exhaust emissions, and field testing, and stands astride science and application. No group of people can better "see" what happens inside a diesel engine combustion chamber (except, possibly, viewers of the fascinating computer-animated multimedia show that introduces visitors to the Diesel Technology Center). Another group, under Sam

Though simulation can help researchers "see" inside an engine, it is often better to actually do it.

Laser measurements, seen through a liner port, produce an eerie but beautiful light.

Barrow (an Englishman living in Switzerland too long to be called English) supervises the study of engines in the field equipped with sensors that monitor thousands of hours of operations. There is a group led by Dominique Jacquet dedicated to helping engines better handle fuels, and who also work closely with major oil companies such as Mobil, Shell, and Exxon, to improve crankcase and cylinder lubricants in Sulzer's general programme of extending time between engine overhauls. Another group, under the leadership of an electrical engineer, Stefan Fankhauser, is developing the "intelligent" management systems that will make diesel engines run more proficiently into the millennium (discussed in great detail later).

One continuing task the Technology Department confronts is meeting the demand for increased engine power (a demand that, although it leads

Sensors and wires on a cylinder cover in development tests of the RTA84T.

to increased knowledge, also leads to increasing costs and human stress). The classic, conservative, way to power up an engine for the market is to take an earlier model and increase its output. But researchers rarely allow themselves the luxury of thinking they will someday reach the limit in combustion pressures and engine powers (30 years ago engines had a maximum pressure of 80 bar, now they run up to 150 bar).

The limits have not been reached. So there is a certain confidence that the size of series engines can go up, because there is enough experience among Winterthur engineers (and confidence in their computer programmes) to predict performance.

And, of great importance, at the Diesel Technology Center engineers can push test engines into running with pressures and temperatures far beyond those produced in current production models, to simulate life in the future. By running test engines for long periods, they can develop simplified components and new bearing materials, and measure temperatures, pressures, component accelerations and vibrations, to assure customers that the larger engines of the future are also the safer engines of the future.

Another area that serious researchers confront daily is that arcane ritual, tribology, or the art and science of avoiding ring, liner, and bearing wear. In the past, the conquest of destructive wear sometimes required radical solutions. When the first version of the RTA58 was on the drawing board, it was designed to have its piston/liner lubrication applied from inside the piston, rather than through the cylinder liner. The engine was, on paper, one of perfect design, done with incredible logic. Its designer won several awards for his work and for papers he delivered. But it simply didn't work.

That was by no means a unique situation — it is a given that research and failure are two sides of the same coin. (Today's ZA40S and ZA50S four-stroke engines, however, have very successful internal lubrication systems.) Other far-sighted notions have come from the intense life of research and development that have failed to conquer tribology include plasma-coated piston rings, and cylinder liners hardened by the Tungsten-Inert Gas (TIG) system, a process similar to welding. Although computer simulations can be used to avoid most major failures, the road to success is often strewn with bad inventions.

As Kaspar Aeberli has said: "It is a good feeling to plan something, and just find that it works!"

P&O's *Pride of Dover* runs before the famous White Cliffs, powered by her three 14-cylinder Sulzer ZA40S engines.

Fundamentally, one of the Technology Department's overarching missions is to see to the production of perfect Sulzer engines, according to Swiss specification, no matter which licensee builds them. In that light, Technology also must see to quality assurance, in a difficult situation where engine prices are being forced downward by market pressures, forcing licensees to cut manufacturing costs, and use more subcontractors (which puts a potential strain on quality).

And, Technology is trying, without compromise, to reduce the time it takes to bring a new engine to market. (At Wärtsilä NSD they don't call the first production engine of a new model by the name "prototype." The feeling is that the word suggests something you fiddle with to get right, then you make a real one. But the combined theory, calculation, testing and field experience of Wärtsilä NSD is assurance enough that the first engine will be ready to drive a ship or generate power immediately, without the need for a costly prototype.)

The Technology Department for Sulzer engines has an open and wide path to follow, in its quest for today's answers to tomorrow's questions and limitations. One group is quietly looking at possible substitutes for the diesel engine. There are many ideas yet to be tried for certain ship markets such as the United States and the Nordic countries, because of likely more stringent emissions regulations. (There are places ashore in the US with such stringent emissions controls that they even prohibit installation of emergency diesel generators for hospitals.) So, fuel cells are being studied as a means to generate electric power to drive large motors because they are very clean. But as yet, fuel cells cannot be made in multi-megawatt units; their maximum output per unit is only in the range of 50 kW to 100 kW; a ship would need 1,000 cells to produce 50,000 kW for propulsion.

Another theoretical area under general discussion is the future of engine cooling. Today, as current engines are constituted, pistons, cylinder liners and covers have to be cooled to limit strains, pre-combustion, wear, and emissions. But, the technologists logically ask: Why do we have add so much massive and costly paraphernalia to an engine just to cool it? Cooling an engine is a decided waste of the energy of combustion (even if some of the heat drawn away is used for

It may have been the first of its kind, but this Sulzer 11RTA96C was not a prototype.

other purposes, such as heating fuel oil or providing hot showers for the crew). One answer that has been toyed with is the adiabatic engine — a constant-entropy machine totally insulated from the outside world, with no intended net loss of heat, therefore no cooling system. It is, however, a long way from reality. (Such an engine, with ceramic cylinders, was tested by the US military to deny an enemy's ability to detect it through heat-sensing equipment. In theory, all the heat of combustion would have been contained — thus hidden from detection — and used to perform useful work. But, by a curious turn of technological reality, the engine turned out to be *more* readily detected by heat-sensing gear than a normally cooled engine.)

The mechanics of the diesel engine are also an area of careful investigation. Already, to reduce weight, without loss of strength, the finite-element method is being applied to the design of welded structural elements and running gear. Weight reduction of a large diesel engine is significant not because an overweight engine can slow (or sink) a ship, but because it equates to reduced cargo capacity and higher manufacturing costs. Along those lines, new crank mechanism geometries have been explored, to possibly eliminate the connecting rod, but these have not yet succeeded.

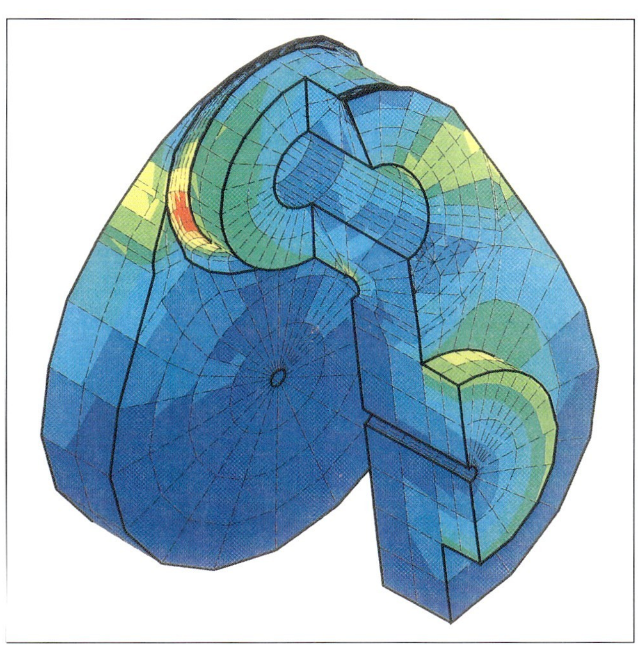

Finite element method, used in mechanical design.

For the future, the Technology group is aware of the enormous potential of the safer, more efficient "intelligent" engine, which will run, monitor, diagnose, and heal itself. And they see the value of new communications technologies that make remote monitoring of this engine routine. Wärtsilä NSD's development engineers are also students of the automotive world, where some of the most exciting engine developments are taking place. They admire the "Bring You Home" concept, in which a damaged engine can re-adjust itself for loss of coolant and continue operating at low power to bring you home. They see promise in the technology of automobile engines with sealed bearings that never need lubrication and sealed gearboxes that never need oil changes.

Technology also sees higher power concentrations made possible by variable fuel injection and exhaust valve timing (already an engine reality), better matched turbochargers with variable geometry that compensate for the difference in charge air required between full and low power (as Alfa Romeo has developed for its diesel automobile). And they dream of a diesel engine with reliability on the order of the aircraft engine's, but greater durability and time between overhauls than today's engine (so they are watching the fate of Porsche's "eternal" car, which will never wear down or, apparently, be embraced by a public that demands a new model every year).

The consummate objective, perhaps, of all the engineers working to improve the Sulzer brand, is the matching of every engine to its ship, by optimising the bore for power, and stroke for speed. Today, any ship can be adequately powered by a choice of several available engines, with various combinations of bore, stroke and cylinder number. Some of the choices, however, turn out to be overpowered by as much as 25 per cent, and are therefore more expensive than need be. It should be quite practical, then, to power a series of standard ships, to produce an absolutely tailor-made engine that is the least expensive possible per ship, as the Japanese have done, with Sulzer engines, in their Future 42 Handymax bulker.

In the ideal world, Kaspar Aeberli, an ingenious man with

a heart dedicated to the future of Sulzer diesel technology, has proposed (with a certain twinkle in his eye) that the company buy its own modest fleet of vessels — container-ships, bulkers, and tankers — perhaps 10 ships in all. After all, he views Wärtsilä NSD as being not just in the engine business, but totally immersed in the business of transporting goods by ship. So why not own a fair sampling of those ships? What better research and development platform would there be than a fleet of active — perhaps even profitable — laboratories on which to perfectly match engines to ships. Is there a "Sulzer Line" ship of the future?

Wherever it may be performed, diesel engine research is not for men in crisply starched white laboratory coats.

R&D for tomorrow

During the great age of sail, grand clipper ships ploughed the seven seas under towering masts and yards draped with billowing sail. Masters drove their ships and crews hard, by day and by night, in calm and storm, through all seasons, to reach port in time to dispose of their cargoes of tea, wool, or nitrates. Always at the mercy of the wind, their greatest concern was choosing the right combination of sails to keep a "bone in her teeth."

Now those great wooden masts and cotton sails have been replaced by a new combinations of "sails": 2,000-tonne diesels driving 10-metre propellers. True, in common with his 19th-century brother, the modern master must also keep that bone in her teeth. But, no 19th-century master ever had to worry about piston performance, cylinder pressures, torsional vibration, engine management, or fuel consumption (unless the "fuel" was a daily ration of that old demon rum). That sort of worry is the purview of diesel engine research and development.

R&D for the Sulzer engine hasn't changed much in its century-long quest for new diesel technology. The essence of research in the design of tomorrow's diesel engines is still to give them larger cylinders and higher pressures, in order to make them more and more powerful.

But the most striking themes of diesel research today are aimed at making engines more reliable, less expensive, and to some extent "foolproof." Most shipowners are professional about running their fleets — they maintain their vessels and equipment properly, navigate them prudently, and keep them in good cosmetic condition. Yet every shipowner is looking for a reasonable and quick return on investment.

Today's VLCCs, for example, have advanced bottom coatings that don't require dry-docking for re-painting more often than three years. If the time between overhauls of their engines could be likewise extended, shipowners are assured that they can keep the vessel in service continually between dry-dock sessions.

Clearly, in order to accommodate shipowners, the next generation of marine engines will be able to operate continuously for 24,000 hours before needing refurbishment. It is an astonishing demand, to be met by research to optimise thermodynamics, engine tuning, tribology and vibration.

At Wärtsilä NSD Switzerland research to attain those goals is under the direction of Nikola Mikulicic. (Mikulici is a Croatian engineer and shipbuilder, and a man of keen mind and sparkling wit, who came to Winterthur to briefly explore an interest in diesels and stayed more than 30 years.) Mikulicic and his 50 researchers (among admiring colleagues called the "Thermodynamic Mafia"), along with the entire Diesel Technology Department in Winterthur, have these main objectives in developing the Sulzer engine:

1) Improving Reliability: A containership carrying US$ 500 million worth of automobile parts for "Just In Time" delivery

Once upon a time, masters would drive ships according to the weather by setting their courses, gallants, royals, and jibs.

Now masters need thousands of "horses" to keep a heavily laden vessel on her proper course in the teeth of an ocean gale.

cannot tolerate engine-related delays; a cruise ship cannot be put out of service even for a single day (its loss of profits from gambling and liquor sales alone would be colossal); a reefer ship carrying California avocados cannot be permitted to arrive late with overripe mush in its holds.

2) Reducing manufacturing costs: Designing engines for cost-saving keeps them competitive, in a cost-pressed world.

3) Improving fuel economy: In recent years, diesel engine fuel consumption has been cut by an astonishing 50 per cent, a unique accomplishment.

4) Reducing emissions: There is a day coming soon when very strict regulations will be in effect.

The research to accomplish these goals always begins from prior experience. Just as shipyards develop new hull lines by incremental improvements in their last set of lines, so does engine R&D begin with existing product. The research team, then, makes a point of following new engines in the field for two to three years to observe their performance and to analyse their operating parameters. The objective is to see all that is right with the engine and, of course, detect what is wrong before it becomes chronic.

The RTA96C

In 1997 alone, three new Sulzer brand engine models passed type tests: two designed for bulkers and tankers (RTA48T and RTA58T) and the Sulzer pride and joy, the RTA96C. From these engines Winterthur technologists are already gleaning enormous field experience, as they have observed their shop tests, attended sea trials of the first ships, sailed on their maiden voyages, and receive steady feedback from the ships.

The development of the RTA96C is a good case illustrating the devotion of Sulzer diesel research and development engineers to their task. Whereas supertankers have long captured the world's imagination (whether delivering millions of tonnes of fuel, or spilling some in accidents), the most interesting and fastest development in ocean-going vessels today is certainly in large, high-speed containerships: In the last decades of this century, containership speeds increased to 26 knots (though they once reached 33 knots in the pre-crisis early 1970s with Sea-Land's SL-7 class ships, which burned fuel quite freely).

In the fledgling days of the RTA series, the early 1980s, the maximum rating of an RTA engine was close to 50,000 horsepower. But by 1988, a 12RTA84 developing 57,000 horsepower was chosen for American President Lines' 25-knot "President Truman" class, 4,340-TEU post-Panamax containerships, the first with breadths exceeding the 32.2-metre Panama Canal limit.

Soon afterwards, the 840-mm RTA84 was upgraded, with an increase in power per cylinder from 4,760 to 5,200 horsepower. The new engine was designated the RTA84C, tailored for containerships of 4000-TEU or more. (The first RTA84C, a nine-cylinder model built by Diesel United, entered service July 1990 in the containership *Katsuragi*.) Soon the RTA84C was in such demand that eight licensees were needed to sell and build them. (Within seven years more than 160 RTA84C engines of six to 12 cylinders were serving more than 20 shipping companies in Europe, Asia, and the US, delivering eight million horsepower.)

By the early 1990s, when the horsepower of the RTA84C had been stretched as far as it would go, it seemed that some sort of limit had been reached. But in 1993, word went around the industry that several shipyards might have orders pending for containerships with capacities of 6,000 TEU or

APL's *President Truman*, here under construction, is powered by a Sulzer 12RTA84.

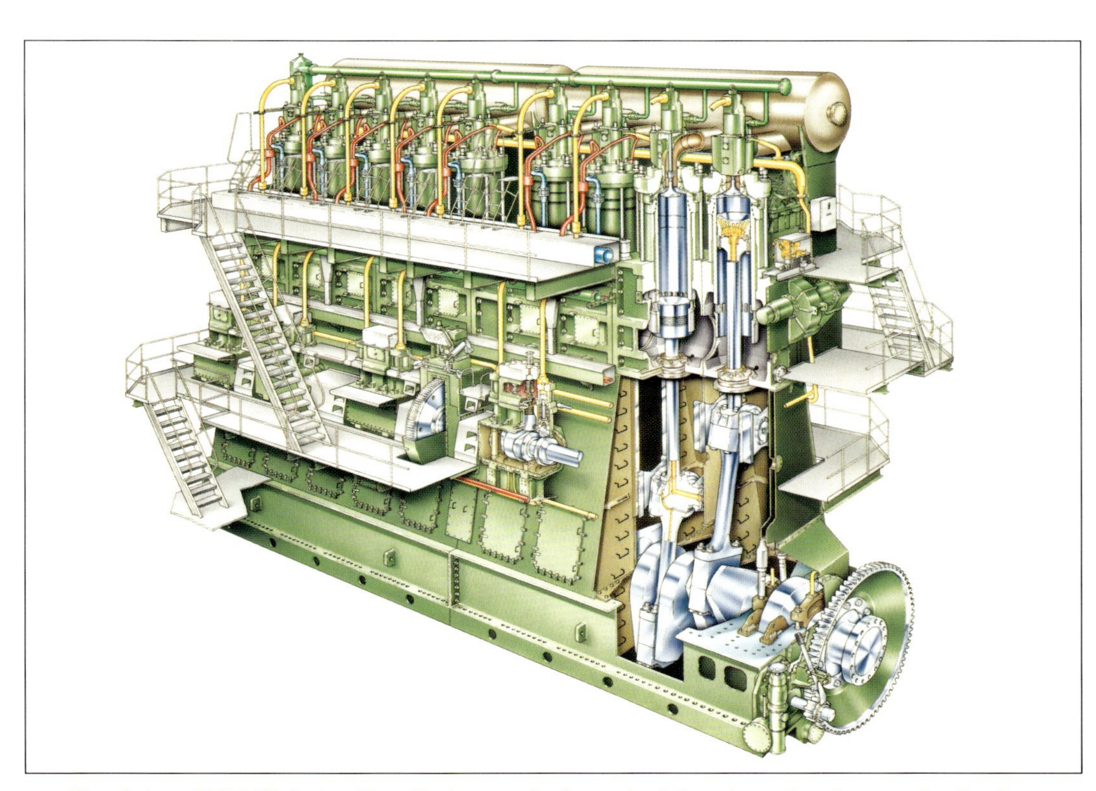

The Sulzer RTA84C (in its 12-cylinder version) reached the pinnacle of power for its day.

more, which would require engines approaching 80,000 to 90,000 horsepower. Analysing just how much horsepower could be put onto one shaft, considering ships with a design draft limit of 13.5 metres and propellers of no more than nine metres diameter, the Research & Development team concluded that though the RTA84C had reached its own power limit, an engine of larger bore was logical. (One element surely driving the analysis was that the MAN B&W 12K90MC, with 900-mm bore, had already delivered 75,000 horsepower.)

In 1993 NSD did a market study and sent inquiries to the containership lines most likely to want such large vessels. Soon NYK ordered two 5,750-TEU ships from Ishikawajima-Harima Heavy Industries in Japan, and P&O Nedlloyd (a merger of P&OCL and Nedlloyd) ordered four 6,674-TEU ships from the same shipyard. Later, Hanjin Shipping ordered two more 5,300-TEU ships from Hanjin Shipyard (in a series that had previously been equipped with MAN B&W engines). All nine ships required new, larger engines; all three shipping lines chose Sulzer brand engines.

In early 1994 the research experts set to work designing an engine to meet the power needs of those ships: a 960-mm bore, 2,500-mm stroke engine. In late 1994 the RTA96C was announced. (Though a 960-mm bore is quite large, just 40 mm short of a metre, the first RND engine of 1968 had a bore of 1,050 mm. It produced 4,000 hp/cylinder; the RTA96C, at almost 7,500 hp/ cylinder, nearly doubled that output.)

Like a champion racehorse being sired by a champion, the RTA96C is a product directly out of its predecessor, Sulzer's RTA84C. At the beginning of the 1990s, researchers had optimised the RTA84C for fuel consumption by persistent trial and error investigations, which included bench tests of some 30 or 40 fuel injector configurations. But the RTA96C's greater cylinder diameter resulted in a lower-aspect, or shallower, combustion chamber than the RTA84C. If engineers had merely copied the RTA84C technology for the RTA96C, the thermodynamics of the larger engine would not have been quite adequate.

The dilemma of a shallow combustion chamber is that it is more difficult to uniformly disperse injected fuel for efficient combustion — fuel may strike the piston and cylinder walls, burn too slowly, create hot spots, increase heat transfer to the cylinder, and produce acidic deposits that can corrode the liner. Thus, with its enormous intended power output, the 960-mm bore presented

***Katsuragi*, the first recipient of an RTA84C (nine-cylinder) engine tailored for large containerships**

researchers with a new level of absorbing, demanding thermodynamic challenges.

Designing for such a critical factor as a combustion pattern is no longer a case of applying experience and intuition, and bench testing dozens of models until the optimal one appears. Instead of applying the tedious trial-and-error approach, engineers went to their computers and simulated a number of fuel-injection solutions using sophisticated thermodynamic software. They checked all of the most promising solutions by calculation, so as to programme the minimum number of highly focused bench tests to make the final choice. (It is similar to the naval architect who does not dare perform towing tank tests on a large number of new hull models, but reduces a new hull design to one or two most promising variations, knowing that building models and towing them in tanks is a rather expensive proposition.)

Computer simulations of the RTA96C cylinder clearly depicted the heat distribution for a wide variety of injection configurations (by showing graphically how hot the cylinder cover and exhaust valve would get, and how various injectors would affect the combustion pattern). The simulations suggested that the number of nozzles be increased from two in the RTA84C, to three (the RTA84C was originally configured with four, but that was found to be too expensive). With the most promising solutions in hand, the engineers designed a set of corresponding fuel injectors to optimise combustion for the RTA96C. Then they took the new injectors to Diesel United, which was to build the first engine. Although a test programme had been planned to choose from among several most promising alternatives, the first test proved that the technology people had hit the nail on the head. (It is a

Temperature profiles are part of the research into combustion chamber design.

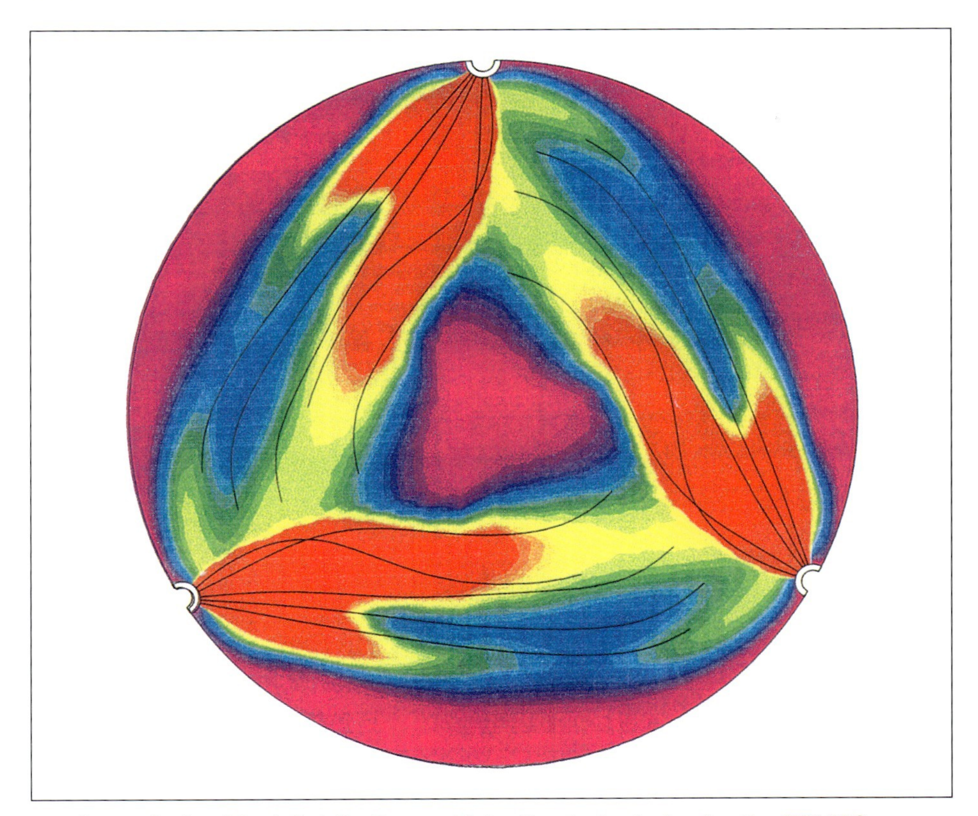

An analysis of fuel distribution and injection trajectories for the RTA96C.

given that engineers *always* know the answer beforehand; they make tests only to *prove* themselves correct.) The three-nozzle format and the injector-hole configuration, along with optimised engine tuning and better-matched turbocharging (at full load each turbocharger gulps in an astonishing 37m³/sec of air at 8,000 rpm) proved the RTA96C to be a thermodynamic as well as mechanical triumph.

The time frame of the development of the RTA96C is also instructive: The first announcement was made in December 1994. In July 1996 the first order was confirmed. On 20 March 1997 the 11-cylinder RTA96C began its shop trials at Diesel United. On the first day, after a few adjustments, technicians "pushed the button" and in 12 to 15 hours it was running at unprecedented power, with good piston temperatures, fuel consumption, exhaust valve temperatures, and little smoke.

The first 12RTA96C for *P&O Nedlloyd Southampton* developed nearly 100,000 horsepower during its shop tests at Diesel United.

In order to gather the most complete data and develop the most complete picture of the engine's operation, during trials of the first production engine engineers installed almost 1,000 sensors — more than 300 thermocouples in and around the cylinders, and some 600 strain gauges. Nearly 30 thermocouples alone were placed in the prime test cylinder, to read *absolute* combustion temperatures at all loads and, of equal importance, to confirm the simulation that, with the three-nozzle configuration, temperature *differences* around the combustion area were within the designed limits as well. All these huge installations were done by Wärtsilä NSD Switzerland's own R&D Development Testing Department, lead by Rudolf Jindra, together with the research engineers from Diesel United.

That engine was in the test bed for two months, running for a total of 161 hours during that period. Initially, temperatures and pressures are measured for two to four days, to determine cylinder performance and blow-by. Stress measurements are taken over 10 days. Checks are continually run on movement of the crankshaft in its bearings, vibration, temperatures, pressures. During the final test, over two days, the engine was again run from 10 per cent to 100 per cent CMCR for the test team to make final adjustments.

The only significant test result that fell out of specifications for the first RTA96C was the Crash Reverse. Because the exhaust valve opening and closing were designed asymmetrically to the BDC piston position, to optimise fuel economy, cylinder temperatures rose beyond expectation during the test and the engine created some smoke. Although this is not critical to safety or efficiency (crash reverses are required for classification, but are rarely used in real life) adjustments were made for future production.

Once the critical tests were completed, the engine was run for 40 more hours to prepare it for delivery to the IHI shipyard. (Unlike the Winterthur Diesel Technology Center, which can run tests with heavy oil, Japan's environmental laws forbid Diesel United from using heavy oil, so this great machine, conceived to burn the meanest dregs of the petroleum cracking process, did not see heavy oil until it was installed in the containership *NYK Antares*.)

The first 12RTA96C — at the time the world's most powerful diesel — was tested at Diesel United in the autumn of 1997. Like its 11-cylinder predecessor, the 12RTA96C passed thermodynamic muster, producing minimal vibration and no rocking, both important to keeping shipowners and ship engineers happy. Although this engine is nominally rated at 89,640 horsepower at 100 rpm, it actually approached 100,000

Seen lying alongside the outfitting quay at IHI Kure shipyard, *NYK Antares* is the first vessel ever powered by a Sulzer 11RTA96C engine.

P&O Nedlloyd

horsepower when, during the final test, it was run briefly at 101.4 rpm. The RTA96C would appear, to ordinary mortals, as a culmination. But it is not seen that way at all to the scientists and engineers who worked two years to create it. As Nikola Mikulicic says: "The RTA96C is no big jump in technology. It's just a big engine." (It is so big that the 12RTA96C can drink 11 tonnes of fuel per hour.)

With the importance of the RTA96C, the design team spent considerable time on combustion simulation, and on fuel and lubricant studies. During 200 test hours in which the engine was tweaked and optimised, Winterthur invested some four million Swiss francs. As a result, the RTA96C is very likely to take more than the normal four years for its payback. But, as the engine has already proved efficient and user-friendly, and it is a prestigious product, its marketing value is incalculable. It has put the Sulzer brand in the public eye as no lesser engine could. This engine was born from a pooling of theoretical design, research, and technology. As a result, there are now more powerfully driven vessels than ever roving the high seas, beginning with the *P&O Nedlloyd Southampton* (pictured left).

Emissions

There is rubbish coming out of oil refineries, and it is consciously aimed right at the diesel market — the diesel engine is the very model of a modern incinerator for disposing of the sludge at the bottom of the barrel. But, there are side effects of this benevolent incineration: pollution. For a long time in this highly industrialised 20th century there was no tendency to curb ship exhaust emissions, except on passenger vessels, where black smoke and white dresses don't mix very happily. Slowly, a "green" groundswell began, to clean up the emissions of automobiles, trucks, power and industrial plants and land incinerators, many of which now have catalytic converters and scrubbers. Since the beginning of 1980s, in the US, restrictions have already been put on ship emissions (vessels that do business on the Mississippi River, for example, are governed by the stricter land regulations). Still, ship pollution is relatively small, when considered against the enormous tonnage of goods ships deliver.

In fact, after six years of work, in the autumn of 1997 the International Maritime Organization agreed to a protocol that limits emissions of the oxides of sulphur and nitrogen (commonly called SO_X and NO_X) for ships whose keels are laid after 1 January 2000, sets stricter limits in the Baltic Sea, and generally restricts emissions of ozone-depleting substances. As a result, there are many promising projects to reduce pollution from ships at modest cost, and shipowners and oil producers have jumped on the bandwagon. Broadly speaking, the two methods of

When testing emissions to protect the world, the test staff must be protected as well.

lowering emission are: eliminate the cause on the input side, or remove its result on the output side. On the input side, oil companies can help reduce SO_X by reducing the sulphur in diesel fuel and heavy oil. (The Baltic Sea protocol reduces the sulphur content of heavy oil to be burned by ships from 3.5 per cent to 1.5 per cent.) Reducing the sulphur content of oil is costly, and will raise the price of bunkers. (Oddly, sulphur has a natural lubricating quality, and reduced-sulphur fuel will lose a residual margin of emergency lubrication.)

One method being worked on in the industry to reduce NO_X is the introduction of water injection into the cylinders along with the fuel. Although spraying water into a hot combustion chamber along with fuel is entirely counter-intuitive to the layman, it works: water injection cools the combustion temperature by 50° to 100° C and results in a reduction of NO_X up to 40 per cent. (In some experiments, water in equal amounts to the fuel has reduced NO_X up to 50 per cent.) The system has been tested, but no ships are using it because they don't yet need such a steep reduction, can't carry or desalinate water sufficiently, and because a water-injection system will raise the cost of engine ancillaries by 10 to 15 per cent.

The dual nozzle developed in Winterthur to inject fuel and water without a costly premix.

The other commonly proposed method of emissions control is the after-treatment of engine exhaust by the addition of catalyzers. Combined with water injection, after treatment will someday reduce current emissions by nearly 100 per cent, but will not appear on ships until they are required by law. Catalyzers, unfortunately, are also costly, adding as much as one third to an engine's price. Reduction in soot output from ships' stacks, however, can be accomplished today by well-matched turbochargers and fuel injectors, and the practise aboard ship of optimising engine load-up.

There is a curious reality to the diesel engine's development during the past 100 years: No matter how far it is pushed, there seems to be room for improvement in all areas. The gradient of the diesel's technological development over time has always been a rising curve — and it will continue to rise, perhaps approaching some distant asymptote, in what seems to be a never-ending process of optimisation. In the spirit of research that sees no boundaries, no limits, Nikola Mikulicic says, "I'm sure we will come to new frontiers, develop new materials, new lubrication methods. We will find better ways to reduce emissions without catalytic converters. We cannot *double* the diesel's efficiency in one blow, but we can add a few per cent at a time.

"Yesterday, solutions could come from just a genius of an engineer. Today, we have very intelligent engineers, but they aren't trained to work on guess and intuition. They use their genius to go beyond science. Our greater possibilities to simulate and calculate today allow us to explore outside known fields, to search for something we never thought about before.

"Our young engineers have an opportunity to deal with all sorts of problems, to learn that they will be happy here. It's not just a job; they can be involved in diesel technology for years from now. But they must be prepared to live with the whole world, to work with all kinds of people, to drink odd drinks and eat snakes if they have to. This is the great enrichment — to work with all people of the world, and never to ask what it is that you're eating!"

Working on all cylinders

Since its commercial introduction in the early years of the 20th century, the two-stroke diesel engine has proven to be the simplest, most reliable, most cost-effective prime mover ever devised for ships. It has relatively few moving parts. It is easy to maintain. It is safe to operate. It is environmentally friendly. And, thanks to the cooperative effort between licensees, shipowners, and the research engineers and scientists at Wärtsilä NSD, it has an ever-expanding life cycle between overhauls.

The basic reason for the success of the diesel engine is its thermal efficiency. There is no comparison between the diesel and that mighty dinosaur, the steam engine. (Still, it is a curiosity of language and history that we refer to a ship under way, being driven by a diesel engine, as "steaming" or "sailing.") But, even as the diesel engine is well established as a prime mover, if one looks at the modern diesel with a keenly sceptical eye, it becomes very difficult to understand how and why it should run. It is amazing that nature and technology can be combined to produce so large and ponderous an iron machine that will turn a heavy bronze propeller at just the right speed to bring a ship on time, to port after port. Year after year.

After all, a diesel engine is nothing more than a great line of cylindrical ovens. The temperature in those ovens rises to 2,500° C, twice a second or more, under a pressure of 150 bar or more, then it cools substantially in the same time. It would seem impossible for the engine to run, in particular because one end of each oven is always moving, up and down, enlarging and shrinking the oven at very high frequencies, in very short periods of time. Worse yet, the moving side of the oven is rimmed with iron rings, and the oven walls are made of iron. They rub against each other constantly, at an average of eight or nine metres per second, and a maximum of twice that speed, with a minimal film of oil between them — barely enough oil in all that heat and pressure to keep the two iron surfaces from abrading themselves into powder in no time.

But the diesel engine *does* work. It is in fact a very good way to use high temperature and pressure for very short, intermittent periods — its key advantage is that its materials don't have time to "feel" the 2500° C temperature. The problem with the diesel engine, however, is that most of its mass and all of its parts are devoted to creating reciprocating motion and converting it into

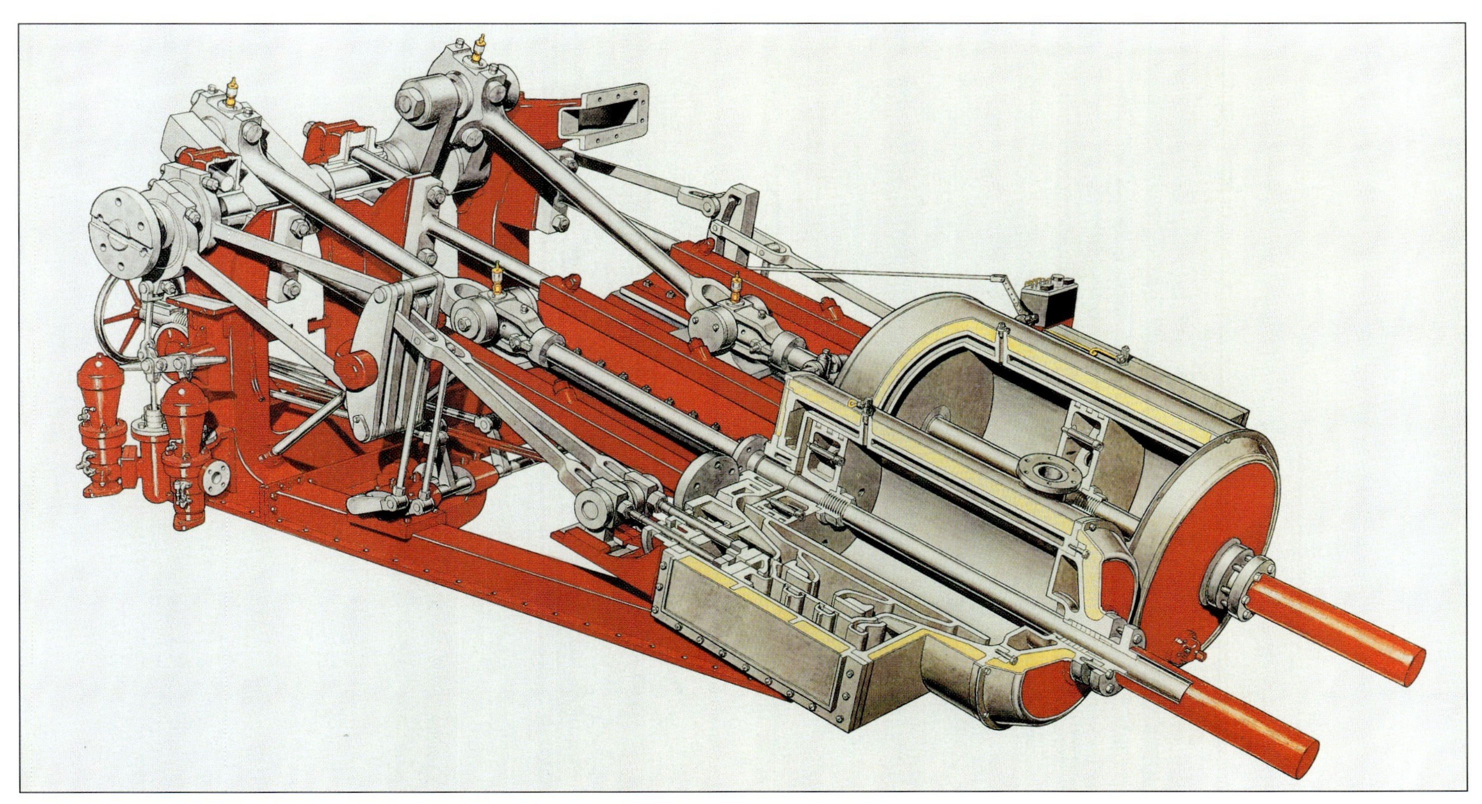

The Sulzer steam engine installed in the early paddle steamer *Pilatus*.

rotary motion — the mind boggles at the inefficiency of such an idea, of such a preposterous system. Imagine, eight, ten, or twelve pistons, rubbing up and down within their high-pressure, overheated ovens, just to force an enormous 600-tonne forged crankshaft to turn at a mere 100 revolutions per minute. It makes no sense.

So the big question is: Will Rudolf Diesel's simple heat engine, based tentatively on Carnot's cycle and permanently on the compression-ignition theory, eventually fade out? Will something replace the diesel engine someday? After 100 years of operation, isn't there some technology, some wild-eyed invention, old or new, that will replace the diesel engine?

There isn't.

High-pressure steam turbines? They are too complicated; they have to create the pressure in one place and move it to another, and their efficiency can never equal the diesel's. Gas turbines? They have the advantage of producing direct rotational power, taking up less space, and costing less. But, they cannot cope with such bad fuel as the diesel, and their parts have a short life owing to their constant high temperatures. Nuclear reactors? Years ago, many young engineers went into the "romantic" nuclear field, but not much has been seen or heard from the few nuclear-power ships built. Fuel cells? These remarkable energy providers — which can burn methane with simple, non-polluting, by-products of CO_2 and water — have much promise for producing heat and electricity in the world of rocketry, cars, and submarines, but are a long way from providing the horsepower demanded by a large ship. Sail-assisted propulsion? With winds blowing in all directions, how many of those gargantuan wind vanes have you seen on the high seas lately? All these classic or futuristic systems have been tried,

©YPS HAMBURG-GERMANY

Despite some of its seeming absurdities, there is nothing available or promised to replace the diesel engine.

but as of today and into our immediate tomorrows, they have not lived up to their inventors' promise. They each have something to offer, but none in total is as good as the good old diesel.

Diesel technology is a very mature, well-established technology. As a result, to the outsider, it looks like a very old-fashioned technology. Yet, thermodynamically, it's a very demanding, very modern technology, because there are no constant conditions within the cylinder of a working diesel — that reciprocating oven again. But, if diesel technology is an old-fashioned technology, it is an amazing technology, because the diesel engine can burn what is left in the cracking process after all the good stuff has been refined, taken away, and sold to the highest bidder.

In fact, a diesel engine can burn practically anything — coal, alcohol, wood, household waste — practically anything (probably) except uranium. Twenty years ago experiments were begun for diesel engines to burn finely-powdered coal in a slurry, using oil to ignite it. Wood slurry (which has no sulphur) has been tried as a means to keeping emissions down. There are still explorations going on to solve the problems of burning coal or wood in a diesel engine. But it's all experimental, so, accepting its fate, the entire industry continues to work to improve the diesel engine's ability to burn typical heavy bunkers, whose quality shows no signs of getting any better.

Still, even with such a poor quality fuel, the diesel process is very hard to beat. All the forecasts say the Earth will not run out of oil, not for a very long time. But, even if it did, engineers would find ways to run diesels into the next millennium with artificial liquids or gases. It will be expensive. But, confronted with that expense, the industry will be inspired to

The 31,000-dwt *Aqua City*, the world's first ocean-going vessel with sail-assist, was powered by a 8,300-hp Sulzer 6RTA58.

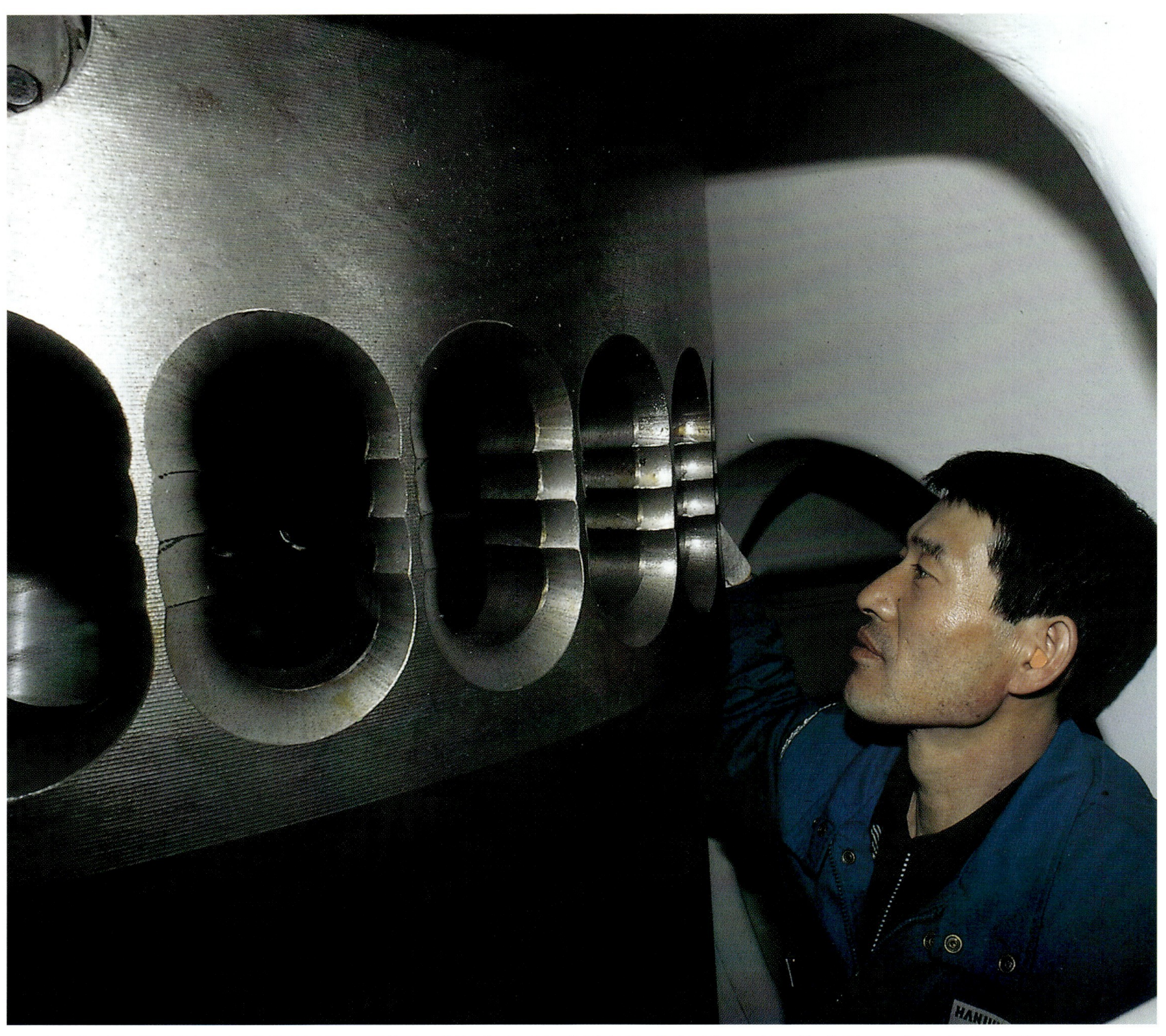

As elegant and basic as it appears, a cylinder liner has complexities ranging from the tribologic to the microscopic.

find ways to make artificial fuel use more efficient, to make it less expensive once more. So reciprocating engines will be mounted in ships for a very long time to come. They stand on very good basic principles.

With ever-higher combustion temperatures and pressures (in particular since the upgrading of the RTA84C engine and the birth of the RTA96C) the demand to reduce piston ring wear and liner scuffing has taken on greater urgency. Ceramic-coated combustion chambers have been built that can withstand these higher temperatures, and the Japanese have a joint venture engine that uses this technology.

But, why introduce the problem of coating iron or steel with a material that might peel off? Better to design components to withstand those high temperatures and the quick cooling, *without* ceramics, by doing what we do today, but doing it better. Materials and methods still have room for improvement — casting, iron composition, machined surfaces, mechanics, tolerances, lubricating oils, fuels, and above all the consistency of components, can all be improved to conquer that old bugaboo, tribology. (Under the frightful circumstances within a diesel cylinder, liners under enormous heat and pressure wear at an astonishingly slow rate — less than 0.1 mm per 1,000 hours — and should last 60,000 hours; piston rings wear less than 0.2 mm per 1,000 hours, with life expectancy of better than 15,000 hours.)

But any time a rare occurrence of excessive engine wear is reported back to Winterthur, the pressure on the designers to conquer tribology is escalated to a fever pitch. (For example, *Zim China,* the last in a series of large containerships built by HDW Shipyard for the Israeli line Zim, and launched in 1997, developed unusual liner wear in her seven-cylinder RTA84C engine in a surprisingly short period after her delivery. The trouble was traced by the field service department to faults in the manufacture of the cylinder liners, and they were quickly replaced by "the Sulzer men," in harbours of opportunity along her route. So tribology is far from being fully conquered.)

There is much work to be done on the hydromechanics of the propulsion of ships through the water, to make better use of the diesel's enormous, efficient power. Waterjets have been tried, but they have relatively

***Zim U.S.A.* takes cargo in the night. She was not bothered by the unique liner problem suffered by her sistership *Zim China*.**

low efficiency except at high ship speeds. It still seems that a large propeller that moves a huge quantity of water during each revolution is unbeatable. (Waterjets have found great suitability, however, in high-speed ferries.)

Demonstration models have been made and tested that convert the reciprocating motion of an engine into the reciprocating motion of a fish or dolphin tail to propel a vessel; there is good sense in not bothering to convert reciprocating motion into rotation, when that motion can (theoretically) drive a ship. This concept is called, by a slightly political sounding slogan, "No More Propeller." But, it will not be easy to replace the propeller, which can only get better.

There are even pie-in-the sky dreams of ionising water and accelerating it as a means to propel a vessel, but until that is worked out the diesel will still do the real propulsion through a propeller.

With the history of diesel technology, from time to time there have been true breakthroughs: in scavenging, turbocharging, or electronic control. But it's still the same basic technology, just with some local refinements. You always come back again to basics. Like steam engines, low-speed diesels are just dinosaurs. But they are strong and forward-looking dinosaurs, and they don't show any signs of dying out.

True, if the future of shipping is to be found in some unimagined invention; if container terminals get so choked by goods being delivered by huge ships that transportation methods change drastically — say, to the use of massive pipelines — then the diesel may die its natural death. But that is unlikely. The engine is too predictable, too efficient, too economical. The transoceanic shipping cost of a pair of sneakers from East Asia to its ultimate user is probably around $0.50, no matter the price of the sneakers (and, as any parent knows, many of them are selling for more than $200). But the price of shipping those same sneakers within the borders of Korea, where many are manufactured, is 35 per cent of the production costs, proving that continent-to-continent transport is cheap, thanks to large ships. And large ships will be using large diesel engines for some time to come.

But they will be better engines. Finite element analysis has permitted more precise, and therefore lighter structuring of engine bedplates, columns, crankshafts, journals, bearings and the like. Software has been developed that allows keener analysis of the sensitive interaction between a crankshaft and its bearings: A crankshaft, with its complex shape and odd pattern of inertia, is constantly pulling away from the bearings, at the same time the firing of each cylinder puts load on the bearings. The structure deforms in a very complex manner. With this new software, the shafts will have better form and the bearings will give long lasting, better support.

Since 1903, when Sulzer set up its diesel research department, with its first order of 12 engines, the company has built a great number of research engines to test old theories and new ideas. Clearly, there are still many tricks left to apply to the old engine concept. But just as clearly, the most significant trick, the most wondrous trick of them all, is the creation of the "Intelligent Engine".

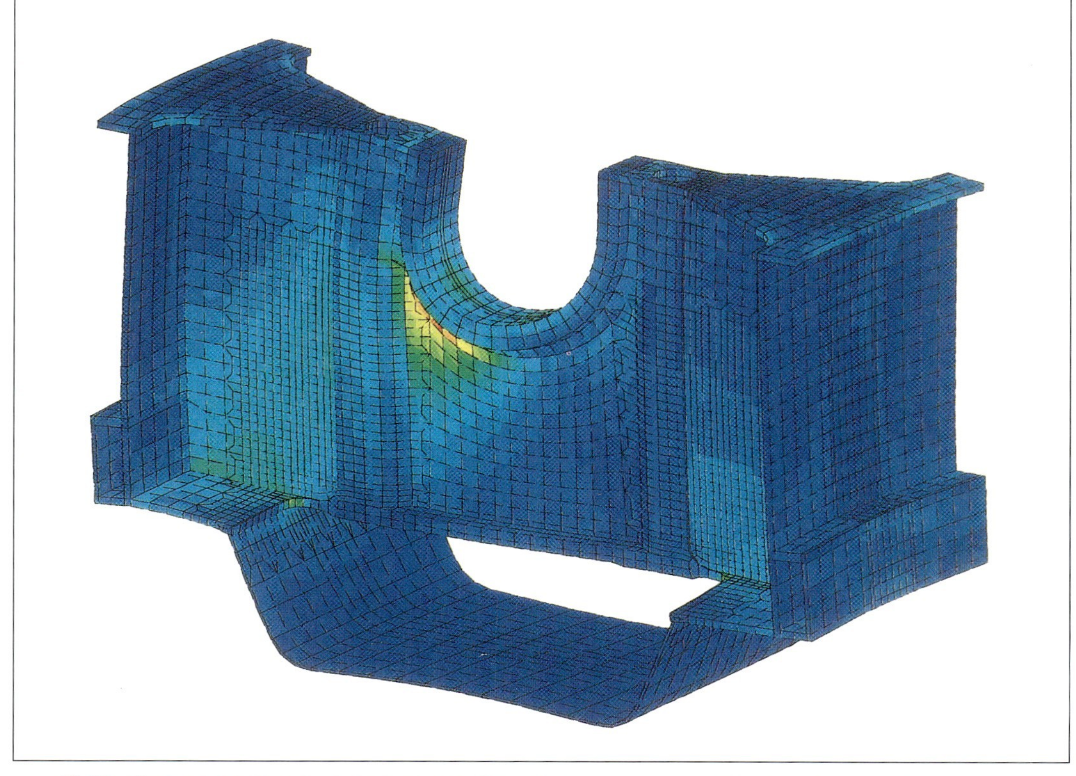

Finite element analysis reveals stress distribution in a bedplate housing bearing.

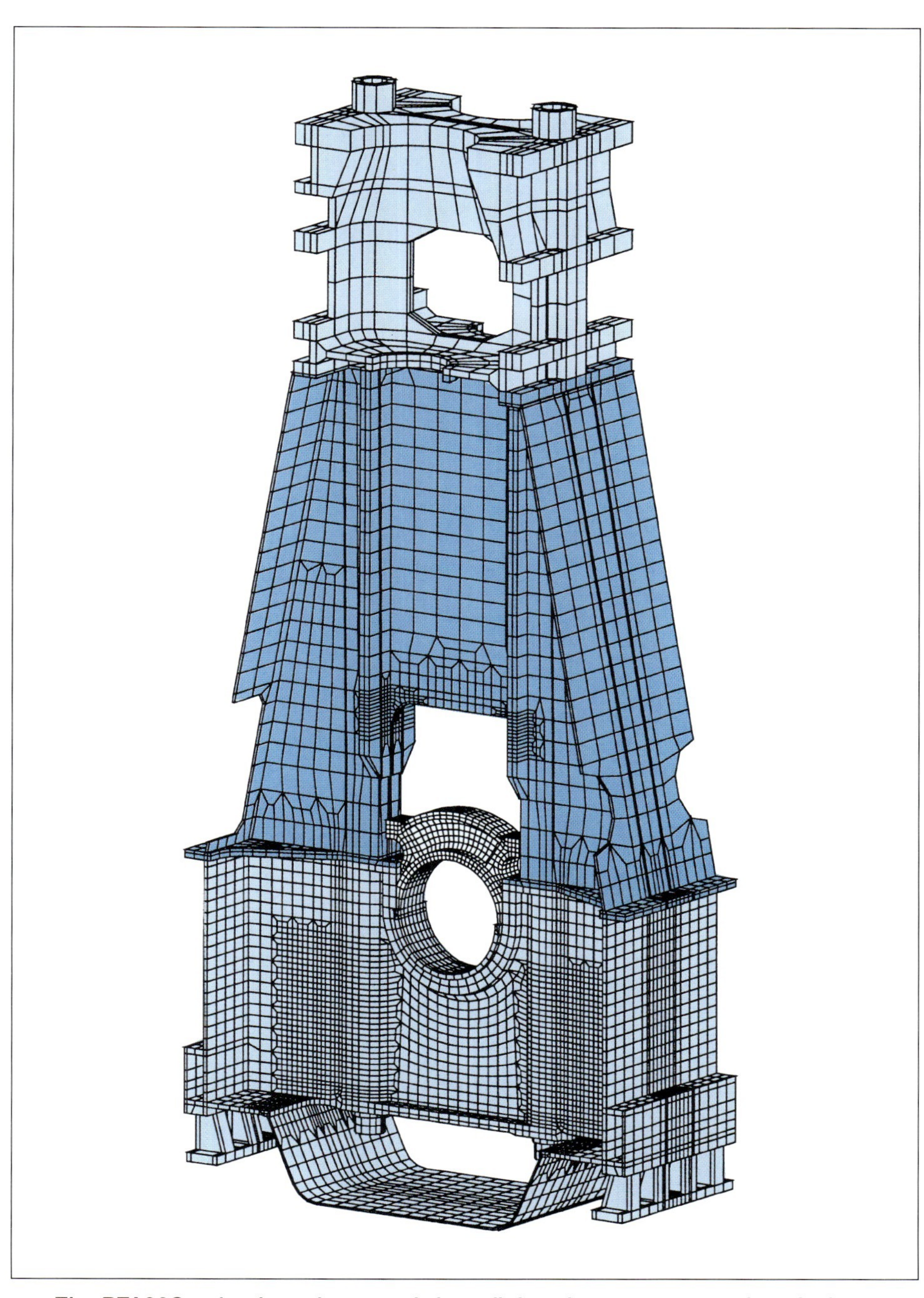

The RTA96C gained much strength from finite-element structural analysis.

The intelligent engine

The intelligent engine is no diaphanous "blue sky" machine; it sits today on the solid foundation of existing technology. It is real. (Nearly a decade ago Sulzer's engineers developed the RTX engine, with completely electronically controlled hydraulic fuel and valve actuator pumps. It still had a camshaft, but it wasn't used.) The intelligent engine is one that "thinks." Its whole operation is governed by computers; it is the ideal marriage between the mighty machine and the minuscule microchip.

For years we have had engines with electronically controlled variable exhaust-valve closing (VEC) and variable injection timing (VIT), functions governed by microchips. Engines with VEC and VIT are already somewhat intelligent: they run more efficiently over the range of their operating speed and loading — one of the diesel's historic weaknesses. The industry still has to prove to shipowners, however, that an electronically controlled engine will not break down as a result of failed circuits, and has comforting redundancy. Then we will see the optimal engine, made optimal for fuel consumption, emissions, or whatever else is pressing at the time.

But intelligence is not just a matter of redundant electronics, it is full engine optimisation. Research has already made it possible for Sulzer engines to run on heavy fuel oils of ever-worsening quality. But the microchip has made it possible for us to talk about intelligence as an overall engine attribute. The intelligent engine is the engine of the future (though many of its components are already here). It is an engine that monitors its own physical state, and adjusts its own parameters for optimal performance in all conditions. When there is trouble, the intelligent engine can move lubricating oil to where it's needed, reduce power in an emergency to control wear and avoid damage, and more.

In this light, Wärtsilä NSD has made a commitment to developing broader, proprietary Engine Management Systems (EMS) that would help optimise engine life by monitoring and analysis of engine performance, maintenance planning, and spare parts control, all essential parts of the "intelligent" engine package.

This change of philosophy began with remote bridge controls. In 1972 Sulzer began producing its own bridge control systems. And until the

Engine monitoring and control are handled today by a touchpad and microchip circuits.

late 1980s, Sulzer manufactured a good portion of the external components needed to run its engines. This was a practical way of assuring shipowners that the "naked" Sulzer engine was "clothed" in the best possible operating mode of the day. But, after a market survey in the late 1980s, Sulzer's management decided that, with so many suppliers producing dependable, competitively priced, systems, the company no longer needed to be in the business of producing remote bridge controls for its engines. So it developed a proprietary technology system to allow only Sulzer-approved remote controls to be connected to its engines, and it began licensing that technology to established manufacturers of remote controls. Wärtsilä NSD calls the system DENIS (Diesel engine interface specification).

The members of the DENIS family are each dedicated to a different type of Sulzer engine, two-stroke or four-stroke, each defining the technical specifications of the engine's interface with Sulzer-approved remote control system. DENIS has no hardware; it is, instead, a software description of signals exchanged between the engine and its remote controls, safety and alarm systems — the engine cannot be remotely controlled without it. The interface technology is licensed to several outstanding companies. Wärtsilä NSD supplies the controls and interface wiring; the shipyard and supplier just hook up the control system.

DENIS not only simplifies control, it allows a shipowner to purchase compatible automation components from one Sulzer-approved source. (Until now, the problem of a multitude of electro-mechanical interfaces supplied by a multitude of manufacturers was a plague on many ships. American President Lines learned that the hard way in the late 1980s when it found that its highly computerised, and highly admired, C-10 post-Panamax containerships developed lots of interface "glitches.")

The second major thrust toward the intelligent Sulzer engine is a system called MAPEX (Monitoring and maintenance performance enhancement with expert knowledge). MAPEX is a set of optional computer-based systems that improve the operating economy and maintainability of a Sulzer engine, and thereby extend its life. The MAPEX family has many "cousins," each dedicated to a specialised function. Together, the family has the means to look inside a Sulzer engine without opening it; to gauge its performance and condition and detect trends in wear, to protect the engine from damage. For example, one member, MAPEX-CR, monitors the key parameters of the combustion process in each cylinder, and sets off alarms if they fall out of specification. MAPEX-PR monitors liner wall, cooling water, and scavenge air temperatures, in relation to engine speed and load, to see that the pistons are running properly. MAPEX-SM, set on the platform of the familiar AMOS-D software, manages spares purchase and inventory, and maintenance history, and can be installed on the ship and the home office. MAPEX-TV warns of excessive torsional vibration to protect flexible couplings, engine component and gears. MAPEX-AV monitors axial vibration.

Since its inception in the late 1970s, SIPWA-TP (Sulzer integrated piston-ring wear-detection arrangement), measures specific radial piston-ring wear and its circumferential pattern in each cylinder. SIPWA is able to detect and measure the slightest ring wear, plot these data as trends, perform diagnoses, and set off alarms when trouble is indicated. By virtue of its ingenious

SIPWA-TP monitors wear through a patented top piston ring and sensor.

(and patented) top piston ring, which signals its own wear pattern to a sensor in the cylinder liner, SIPWA-TP gives the status of piston ring condition, pinpoints the location of the ring wear concentration, and reveals ring rotation patterns. By this system, engineers can optimise lubricating oil distribution for maximum efficiency, take quick countermeasures in case of abnormalities, and avoid unnecessary piston overhauls.

In the long run the shipowner has the potential to reduce lube oil consumption, lower operating costs, while extending time between overhauls. The payback in lube oil savings for the system can be just one to two years. Without SIPWA you cannot know there is serious ring wear until the engine loses power. Above all, because measurements made by SIPWA are at the end of the working chain, they also are a prime tool in determining the cause of the wear earlier in the chain — be it bad fuel, water in the fuel, poor lubrication, etc.

MAPEX needs 20 to 50 hours to calculate a trend, which is adequate for monitoring of two-stroke engines, where (barring disaster) the processes of wear is slow and there is time to react. But with four-stroke engines, running at higher speeds, there is less time to react, so MAPEX-PR uses two cylinder sensors to detect short-term trends, and acts as a true early warning system.

Shipowners who are true professional take care of their ships for the long run, and many of them consider MAPEX not an option, but essential to protecting their investment. Many shipowners have retrofit MAPEX, which can readily be done as all Sulzer engines are ready to accept sensors and monitoring gear. MAPEX systems are sold only through Wärtsilä NSD, and the company shares the technology exclusively with companies that do not manufacture or market competing products.

One of the first SIPWA systems, in the late 1970s, was installed on the P&O containership, *Remeura Bay*. It enabled the shipping line to significantly reduce the cylinder oil feed rate, without added wear, and subsequently the system was installed on other ships of that line. On another ship, *Bellatrix*, SIPWA saved the day. Shipboard engineers were not aware that their bunkers contained an excess of abrasive refinery residues (cat fines), which were not being filtered out by their fuel system. Piston ring wear increased sharply, and MAPEX sent an alarm; within hours the engineers found and fixed the cause and the engine returned to normal.

Bellatrix **engineers learned the value of SIPWA-TP when an alarm warned them that ring wear had increased sharply from cat fines.**

Of course, remote transmission of engine performance data is already in practice. Through a programme Winterthur arranged with BP and Chevron, several of their tankers are transmitting regular reports to their own home offices for analysis. Beyond that, the P&O Nedlloyd containership launched in the spring of 1998, and driven by the first 12RTA96C engine, is transmitting, through its SatCom system, daily reports on engine operation parameters directly to Stefan Fankhauser and his Engine Management Systems Department in Winterthur. In that way, the expert eyes of the people who created the system are supplementing those of the ship's engineers and the owner, to monitor the engine performance. This is clearly the trend of the future. (The trend has also affected Wärtsilä NSD's main competitor, MAN B&W. Unlike Sulzer diesel engines, which have double-valve control of injection, MAN B&W engines have long used Bosch-type fuel pumps with timing controlled by a helix. This allows variable timing only to the end of the injection, but fixes the timing of the start of fuel injection for the engine's life, unless the helix is changed. In early 1998, however, MAN B&W announced that its 6L60MC, with electronic injection and valve timing, would be installed on a Norwegian chemical tanker, raising the ante once more between the two world competitors.)

Another important trend into the future is the very strong, and quite fascinating, prospect of a system rather artlessly called Engine Care. Unlike the human-driven License Care, Engine Care is a highly technological, stunning advance for the future. Engine Care is a system by which the engine (with the help of a computer) will analyse its own problems, and to the extent possible, remedy them on the spot, untouched by human hands. The computer would be "trained" to make constant comparisons of current operating parameters to a comprehensive database of engine operating experience, in order to recognise when engine parameters fall out of their proper range. By case-based reasoning, the computer simply looks for a match of the current situation with any past case that led to a problem or failure. Thus, it will anticipate the failure before it happens, and either set off an alarm or, in the most advanced situation, take remedial action, such as increasing its own lubrication oil feed rate.

With electronic injection, electronic valve timing, DENIS, MAPEX, and Engine Care, the Sulzer brand diesel can expect

All of the President Truman-class APL C-10 containerships were equipped with the SIPWA-TP piston-ring monitoring system.

a long life. Intelligent engines have those electronic controls, but they still have camshafts, followers, push rods, and poppets because, as yet, no shipowner wants to be the first to accept an engine without a camshaft. Shipowners, who can be as traditional as diesel technology itself, cannot yet reconcile a camshaftless engine. They are shipping men: they still believe that the diesel is a rugged machine that should be operated by humble men; they are not willing to hire a Ph.D. in electronics to drive their ships.

But no Ph.D. will be necessary when the completely intelligent engine makes its long-awaited debut. In several years the mechanical and electronic forces of the future will marry and live happily ever after. Then, as confidence builds in the microchip, the old engine will be replaced by the new. It will be an engine with lower fuel consumption, improved tribology, higher power, greater efficiency, and unmatched reliability. It will meet all the future's restrictions on emissions. And it will be able, in many cases, to treat its own ills — or at least have the intelligence to make a call to a specialist doctor, "the Sulzer man," who will prescribe the cure. What more can you ask?

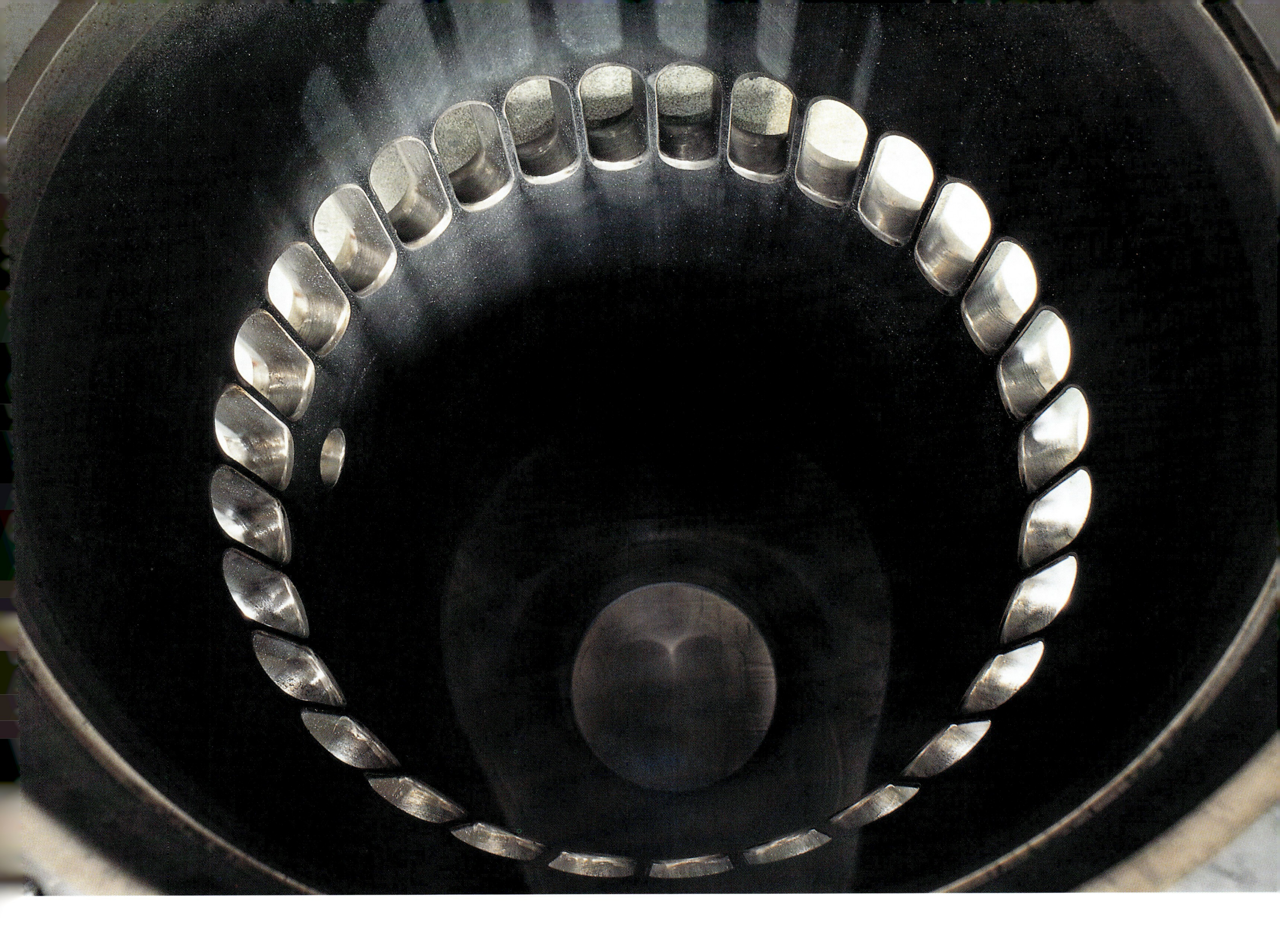

And what of tomorrow?
The Sulzer diesel engine is, in a sense, a floating ambassador to the world. Every time a ship driven by a Sulzer engine exchanges containers, off-loads crude, transfers ore, breaks ice, or takes on passengers, a message is given: From a simple, theoretical notion in the late 19th century, the Sulzer diesel engine evolved into one of the glories of Mankind's inventiveness. It is the workhorse of the world, and the name Sulzer is ever behind it.

There are about 300 shipowners of significance in the world. Among them is a core of about 100 transportation lines that own 80 per cent of the world's significant ships — such lines as Evergreen, P&O Nedlloyd, Hanjin, Carnival, Zim, Hapag Lloyd, NYK, Royal Caribbean, Norasia, HMMC, Grimaldi, APL/NOL, Superfast, Chevron, AP Møller, and BP. These shipping and cruise lines are the heart of the ocean-going world; they are large companies but in every case they are run by real men with such distinguished names as Arison, Wilhelmsen, Chandris, Chang, Lemos, Livanos, Ofer, Sterling, Panagopulos. To these owners, the initial cost of a ship (and consequently its engine) is a large issue they face in

expanding their fleets, because every shipowner is concerned with how soon a new ship's operating revenue repay its initial cost. But the pressure for return on investment is greater today than ever before, as over-capacity in many markets is forcing freight rates down and caused shipping lines to share capacity with partners, or to merge outright. (The cruise business, which expands every day, seems to be an exception.)

Wärtsilä NSD cannot set the price of its licensee-built diesel engines, so the company must promote its product on other terms. This demands close personal cooperation between marketing people and licensees, particularly when a new engine hits the market. Creating a new engine represents an enormous investment for the Winterthur designers: it can cost the Technology Department millions of Swiss Francs. And the licensee selected to build the first engines also has an enormous investment in design, tooling, and labour. Like shipowners, Wärtsilä NSD, too, expects a reasonable return on its investment.

In the late 1990s Sulzer brand licensees were selling between 140 and 160 two-stroke engines per year. One advantage of the merger with

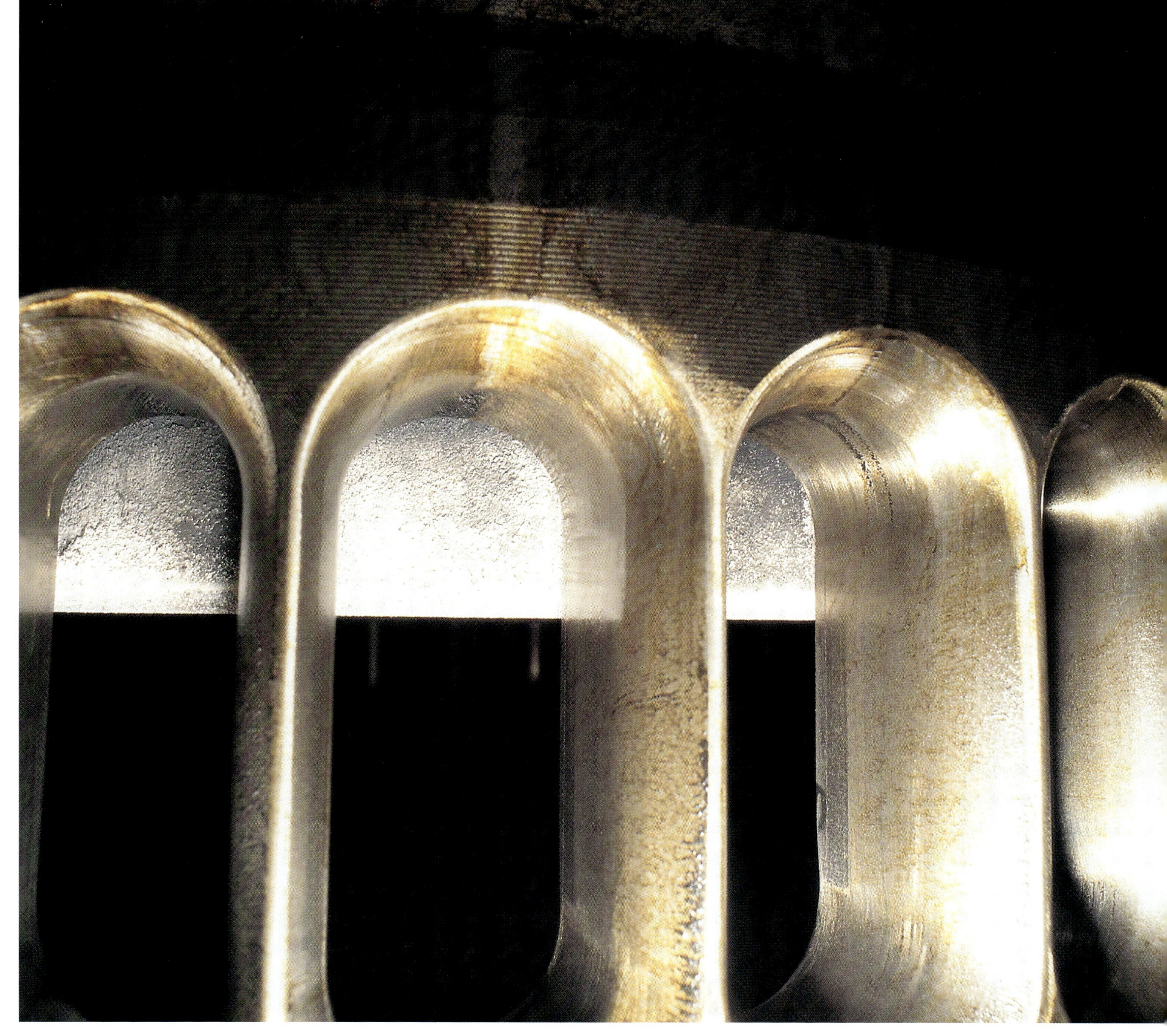

Wärtsilä is that the new technology base, the R&D capability of the new company, makes the Sulzer brand a more potent one for the market. As Christoph Studinka, head of marketing for Sulzer diesels has said, "This is a mature industry with very large cost pressures, but very small technical steps."

But the Sulzer brand still has its work cut out for it. Peter Sulzer, Executive Director of Wärtsilä NSD Corporation and Chairman of Wärtsilä NSD Switzerland, has said: "Over the years, Sulzer has manufactured steam engines, weaving machines, medical equipment, and more. But only the diesel engine has carried the name Sulzer around the world. When people recognise my name, *Sulzer*, they never ask if I come from the same Swiss family that made *weaving machines*. They are thinking of *diesel engines*, and *only* diesel engines."

There is great, justified, exultation in those words. The Sulzer brand, the Sulzer name, the Sulzer reputation are at the very foundations of the global diesel engine industry. And they will continue to be so as long as Swiss reliability, Swiss quality, Swiss caring, and a *soupçon* of Swiss humour remain the firm's strongest assets. But there is much more

to this world than reliability. Peter Sulzer feels that there should be a fair and reasonable living made by *everybody* in the two-stroke diesel business — not just Wärtsilä NSD, but its market rivals, MAN B&W and Mitsubishi, as well. But there is a race going on — the race to push up horsepower, cut prices, and to develop new engines too frequently.

The primary rule of this demanding race is: *If the competition is ahead, we have to do something better*. The result, according to one thoughtful Wärtsilä NSD researcher, is that "we have to design 2,000 tonnes of steel from scratch each time." (After the announcement of the 960-mm RTA96C, for example, MAN B&W announced a larger, 980-mm bore containership engine. And there is subdued talk in the halls of Winterthur that if container harbours would only be made deeper, and propellers would only get better, Sulzer could produce a 14-cylinder RTA96C with 14,000 more horsepower than its 12-cylinder version.)

"In the old days," Peter Sulzer says, "we built the same engines for years, royalties came in, and everybody was happy. In the 1950s and 1960s an engine 'generation' was

around fifteen years, and as many as three thousand engines of a type could be sold in that time. Now a generation is as short as four or five years, with the possibility that only a few hundred engines will be sold, but at higher development expense." (Peter Sulzer also recalls that in the early 1990s, after a brief shipbuilding hiatus, many Greek shipowners returned to the market to build new ships and asked New Sulzer Diesel for its old reliable RND engine. But by then Sulzer was making the not-so-reliable RL engine, and that "progress" almost cost Sulzer the orders.)

And the race for bigger and newer product affects everybody: Shipyards have to constantly redesign engine rooms; licensees have to frequently retool; research and development people have to suffer stressful "teething" under enormous market pressure to bring forth a new engine. It seems to be a natural law that diesel engine development goes in small leaps, as it does with consumer goods where the new model is the key to survival. With the compulsion toward higher power and fewer cylinders, a cynic might conclude that, by some mysterious new law of *increasing* (rather than *diminishing*) returns, the industry's ultimate goal is to produce a zero-cylinder engine of infinite power. In the race for high power and low price someone will cut corners. Then there will be problems.

If the race continues, fewer ships will sail the oceans with the same tailor-made engines; the logistics of maintenance, repair, and spares supply will become overly complex; and engines will be sold at market prices that cannot cover their costs. With such conditions the industry cannot survive.

It is time for cooler heads (even as it is time for cooler pistons) to ease this cycle. There is no new prime mover sailing over the horizon that threatens the diesel engine, that improves upon the diesel engine, that will replace the diesel engine. It is time to let the dust settle for a while so that shipowners — Greek or otherwise — can build more of the same ships with more of the same engines.

Mankind is ever searching for extraterrestrial intelligence. We scan the heavens for rational radio signals that might point to a thought process out there that we can read into, understand, and perhaps eventually communicate with. Now that the intelligent engine is a near reality, we have to search for intelligence here on Earth — intelligence of the diesel engine marketplace. The diesel engine seems, like a healthy child, to be growing in value too fast for its shoes, while its price comes down. It may suffer the slings and arrows of outrageous tribology, but the old workhorse still works. It works beautifully. And it works better all the time. Why not let it work for a while, unchanged.

The diesel engine is an astonishing construct. But it is a construct of the warmth of the human spirit and the heat of the human mind, more than the cold laws of technology. As Heinrich Schneider says, "After one hundred years the brand needs people with vision. Otherwise it will disappear like the old Doxford engine. To my younger colleagues, I say, you have a chance. You must surely take that chance. But it is not free of charge."

Whatever its future holds in the way of high combustion temperatures, mean effective pressures, and explosive brake horsepowers, Rudolf Diesel's invention is here to stay.

And the Sulzer brand diesel engine, under the bold blue-and-orange Wärtsilä NSD banner, will be there to lead the way *...from the mountains to the seas.*

Helmut Behling

From Bremen, Germany, he was educated in engine construction. Serving some years as an engineer on commercial vessels, he encountered his first Sulzer diesel in 1959. Returning to land, he entered the aerospace industry, working as a project salesman for Airbus. In 1989 he started experimenting in marine and industrial photography. His first photo book, published in 1997, documents the final shipbuilding activity of Bremer Vulkan. He uses photo equipment of Gitso and

David T. Brown

English by birth, he graduated in marine engineering from the University of Newcastle upon Tyne in 1970. After training and time at sea as a ship's engineer, he spent 12 years as journalist and editor with a major international marine periodical. He joined Sulzer Brothers Diesel Division in 1983, becoming Manager of Promotion and Press Relations in 1990. His hobby is the history of ships and their machinery, which dovetails well with his research into the history of the Sulzer diesel engine.

Jürgen B. Gerdes

Originally from Haren (Ems), Germany, he is a graduate chief engineer from the Hochschule Bremerhaven. After serving as an engineer on commercial ships for several years, in 1992 he joined New Sulzer Diesel in the applications engineering field. In 1995 he went briefly to Japan, then moved to Korea to build up a more effective technical liaison office with Sulzer engine customers in East Asia. He returned to Winterthur in 1998 to become Manager of Marketing Development for Wärtsilä NSD Switzerland.

Jack A. Somer

Born in New York City, where he was educated in engineering, music and art, he spent 15 years in the music recording industry before turning to sailing as a profession in 1971. After some years as a yacht master and ocean navigator, in 1982 he became a full-time marine journalist, and was appointed Editor of *Yachting* magazine in 1985. He is the author of four books on large yachts and commercial ships: *Juliet*, *The Creation of a Masterpiece*; *Izanami*; *Ocean Giants*; and *Ticonderoga*.

Acknowledgments

When the title of "author" is bestowed upon one name in the production of a complex and sentimental book such as this, it means only that the so-named person has been privileged to act as a conduit for a great flow of information from many sources.

In this case, the sources have been as diverse and rich as for any book I have authored. I wish, therefore, to thank all the individuals and groups who gave the information freely, and those who subsequently helped convert it from an encyclopaedic melange to a work meant to be enlightening, coherent, and, ultimately, entertaining.

First, I should like to thank Christoph Studinka and Jürgen Gerdes, who over a 1997 lunch conceived the idea of celebrating the 100th anniversary of Sulzer Brothers' first diesel engine with this book. Gerdes, in particular, has been tireless in his efforts to conceptualise, organise, and finally bring to fruition this most difficult project. And he never hesitated to roll up his sleeves to help turn a vague, working notion into a clear, lasting reality (and all in an impossibly short time).

I also thank David T. Brown, whose equally tireless efforts for a dozen years (on his own time) to reconstruct more than 100 years of Sulzer diesel technology and history made the first section (and much more) of this book so informative. His thoroughness is found everywhere on these pages (but any abridgments of his work are to be blamed on me). The massive photo research for nearly the entire book must also be credited to Gerdes and Brown.

I want to acknowledge, by as many names as memory permits, the managers, engineers, scientists and friends of Winterthur who allowed me to keep them from their work (and perhaps play) by my interviewing them during two whirlwind weeks in Februrary 1998. They include: Kaspar Aeberli, Peter Baumgartner, Rudolf Demmerle, Stefan Fankhauser, Jürg Flach, Markus Geist, Werner Jungblut, Nikola Mikulicic, Klaus-Dieter Muuss, Wilfried Pernter, Frédéric Porchet, Felix Prochaska, Albert Sennhauser, Leo Schnellmann, Heinrich Schmid, Wolfram Schultz, James Thomson, Martin Wernli, and with extra appreciation, Heinrich Schneider(ovski). Special thanks go to Armin Probst, who cost me the most expensive telephone call of my life. And among retirees from Winterthur I cannot forget Willi Würgler, Gottlieb Wolf, and Jan-Arie "Jackie Typhoon" Smit.

I wish now to thank two men who first approached me to write this book, and who acted as friendly surrogates for Wärtsilä NSD Switzerland: Bruno Hug and Oliver Prange, of DENON Publizistik, Rapperswil, supported, encouraged (and occasionally fed) me during the weeks I wrote and designed the book.

I wish to express appreciation to Freddi Mathys and the wonderful staff of Zürichsee Druckereien, Stäfa, for their unending hospitality during three arduous weeks of work, with particular thanks to Ruedi Züger who scanned hundreds of images. As to the book's layout, thanks must first go to Edith Camen, who created the basic design. But, more important, I am particularly grateful to Franziska Rose and Sherif Ademi, who used the universal languages of intelligence, artistry, and enthusiasm to help me put these pages together with taste, colour, and balance.

Finally, I want to express my most profound thanks to Peter Sulzer. Had I not received his enthusiasm and confidence to probe into his family's brilliant history and into the rich living traditions of the modern Sulzer diesel, its licensees, and its users, none of this would have been possible. I too have become a *"Sulzer man."*

J.A.S.

Appendices

Worldwide Sulzer diesel engine orders of all types

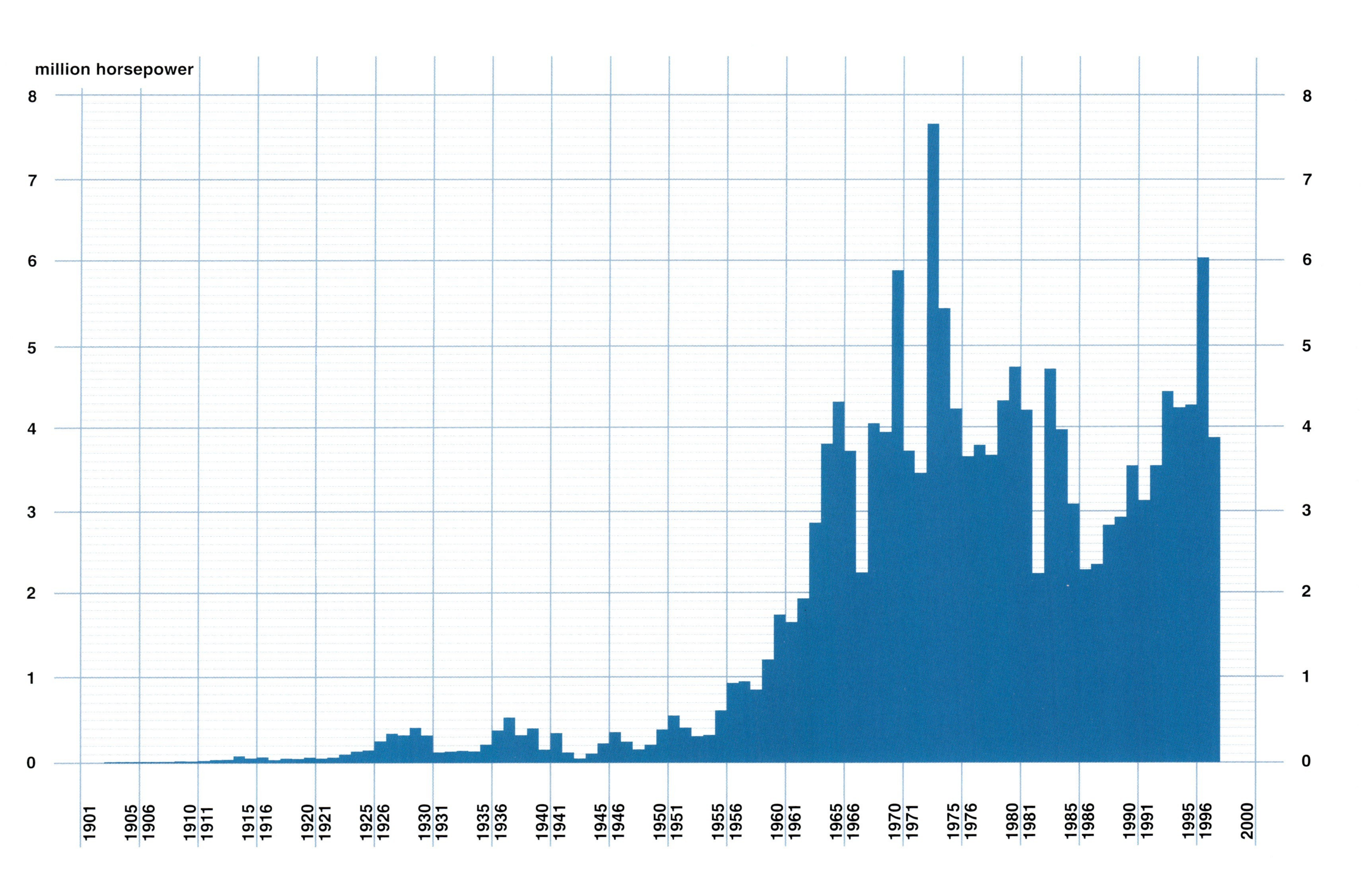

Summary statistics of Sulzer two-stroke diesel engines (end of December 1997)

Years introduced – finally built	Engine family	Typical designations	Bore in cm	Number built & ordered	Total horsepower
Low-speed two-stroke engines					
1908–31	S	SNo., SL, ST	18 – 90	604	1,132,110
1908–32	Z	ZA, ZF, ZPF, ZT	52, 60, 65, 68, 72, 76	147	278,970
1916–31	Busch-Sulzer	Various	Various	743	280,310
1929–33	DZ	DZL, DZD	60, 70, 76	7	49,200
1929–37	SN	SN	48	8	12,000
1930–46	DS	DSD, DSDT, DSDG, MSD, KD	53, 60, 70, 72, 76	36	224,500
1932–64	SD	SD, SDT, SPD, MS, KD, SDS, SDH, SF	36, 40, 48, 49, 56, 58, 60, 65, 68, 72, 76	744	3,638,885
1934–54	ZD	ZD, ZDA	60, 65, 72	10	50,470
1950–63	RS	RSD, RSG, RSAD	58, 76	135	1,086,500
1950–63	SAD	SAD	60, 72	127	675,820
1956–87	R	RD, RF	44, 56, 68, 76, 90	2,100	23 758,300
1967–81	RN	RND, RNF, RNMD	68, 76, 90, 105	2,231	32 629,840
1976–89	RN..M	RND..M, RNF..M	68, 76, 90	607	10,303,040
1977–88	RL	RLA, RLB	56, 66, 76, 90	500	7,185,675
1982–	RTA	RTA, RTA..M, RTA..C, RTA..U, RTA..T	38, 48, 58, 68, 76, 84 52, 62, 72, 96	1,963	40,373,580
	Total			9,962	121,679,200
Medium-speed two-stroke engines					
1912–31	Q	U, Q, QL, QN, M	23 – 65	518	683,720
1916–29	MC	MC	10, 16, 20	162	870
1922–29	RV	RV	20, 24, 25, 27, 31	695	56,990
1925–55	RK	RKP, RKW, RKS, RKH	15, 20, 21, 25, 30, 36, 38	1,900	331,526
1930–37	DQ	DQD	38, 42, 68	6	18,400
1932–42	QD	QDC, QDCT, QPG, QPD, VQA	32, 38, 42, 51, 54, 68	202	425,090
1932–83	T	TS, TD, TG, TSG, TH, TPF, TA, TAD, TMPF	18, 23, 24, 29, 32, 36, 48	2,075	2,963,398
1935–49	ZG	ZG, ZGH, ZGR ZGX	7, 9, 14, 17.5	1,040	50,200
1942–69	M	MPF, MD, MG, MH	32, 42, 51	160	395,040
1967–82	ZH	Z40, ZH40, ZV30	30, 40	99	545,870
	Total			6,857	5,471,104

Summary statistics of Sulzer four-stroke diesel engines (end of December 1997)

Years introduced – finally built	Engine family	Typical designations	Bore in cm	Number built & ordered	Total horsepower
Medium-speed four-stroke engines					
1903–29	D	D, DAc, DAcA	20.5 – 60	1,678	305,548
1910–20	H	HC	12 – 31	55	3,020
1910–28	K	K, KAc	28 – 64	265	92,590
1912–77	LD	LDA, LTD, LHD, LF, LFA, LAH, LV, LD	19 – 34	3,605	5,481,875
1923–28	KS	KS	31	15	2,250
1928–29	DH	DH	38	19	10,700
1928–36	DN	DN, DNAc	38, 45, 54, 60	48	39,500
1928–37	DV	DV	31, 38, 48	54	21,140
1930–63	DD	DDP, DDA, DDH, DDM, DGa, DDW, DDAW, DDAH	22 – 65	1,122	385,825
1934–46	KD	KDP, KDA, KDAc, KV	22–60	80	34,695
1943–76	B	BA, BW, BF, BAH, BAF, BR	22, 29, 36	3,933	1,682,722
1960–72	LVA	LVA	24	109	327,000
1962–71	SLM	Total built after takeover by Sulzer		743	218,016
1968–	A	AL, AV, AL..H, AL..R	20, 25	4,870	4,200,523
1971–81	52/55	V52/55	52	2	25,320
1973–75	65/65	V65/65	65	1	21,600
1973–	AS	ASL, ASV, ASL..H	25	706	1,259,545
1973–	Z40	ZVB40, ZL40, ZV40	40	894	5,923,367
1982–	AT	ATL, ATV, ATL..H, ATL..R, ATL..GL	25, 27.5	589	1,006,442
1984–88	ZA40	ZAL40, ZAV40	40	28	201,840
1985–	ZA40S	ZAL40S, ZAV40S, ZA40SG	40	738	8,966,040
1988–	S20	S20, S20U	20	601	869,828
1995–	ZA50S	ZAL50S	50	6	88,020
	Total			20,161	31,167,406
	Grand Total	**Sulzer Diesel Engines as of end of December, 1997**		**36,980**	**158,317,710**

Chronology of Sulzer two-stroke diesel engine developments

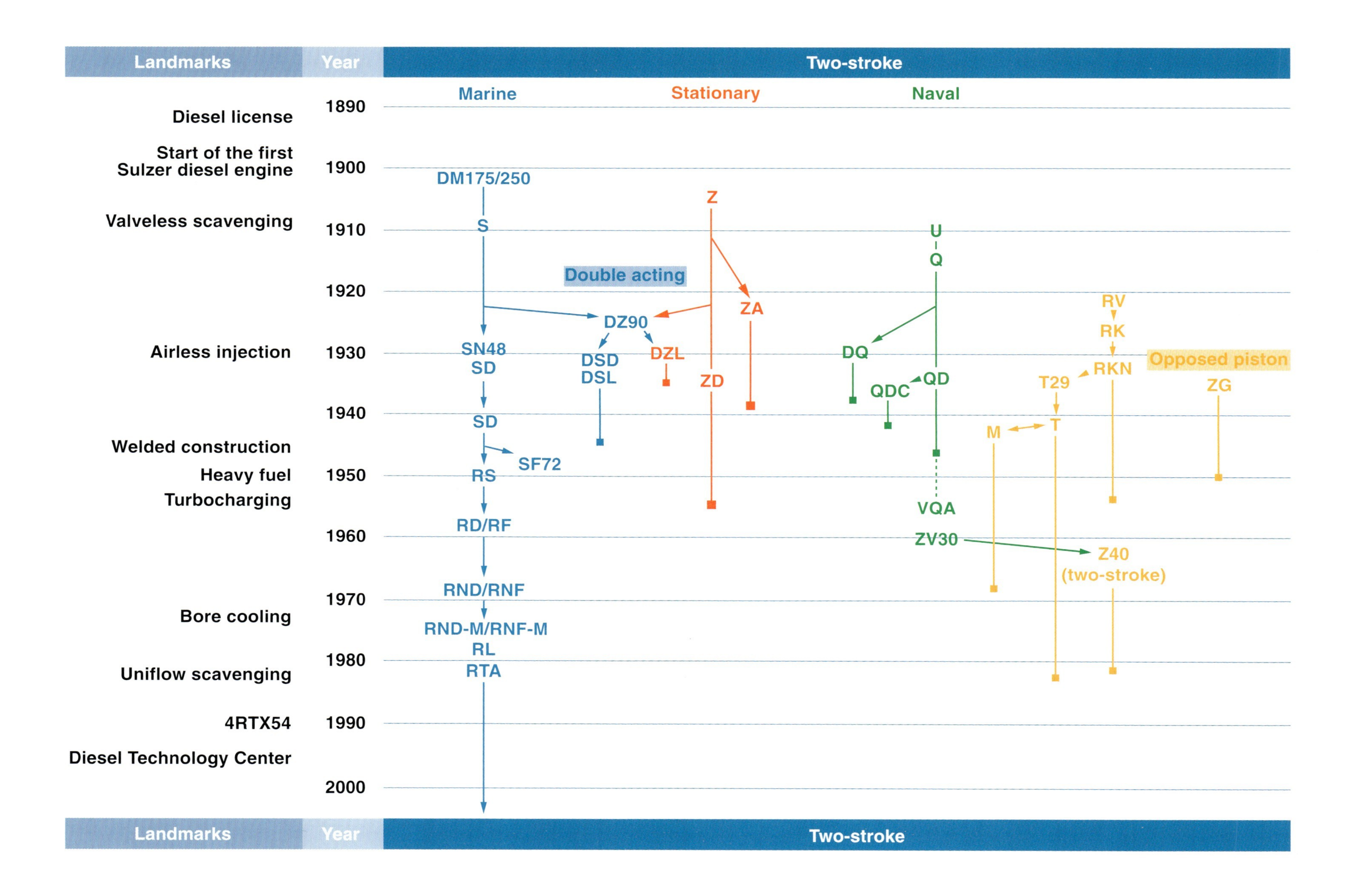

Chronology of Sulzer four-stroke diesel engine developments

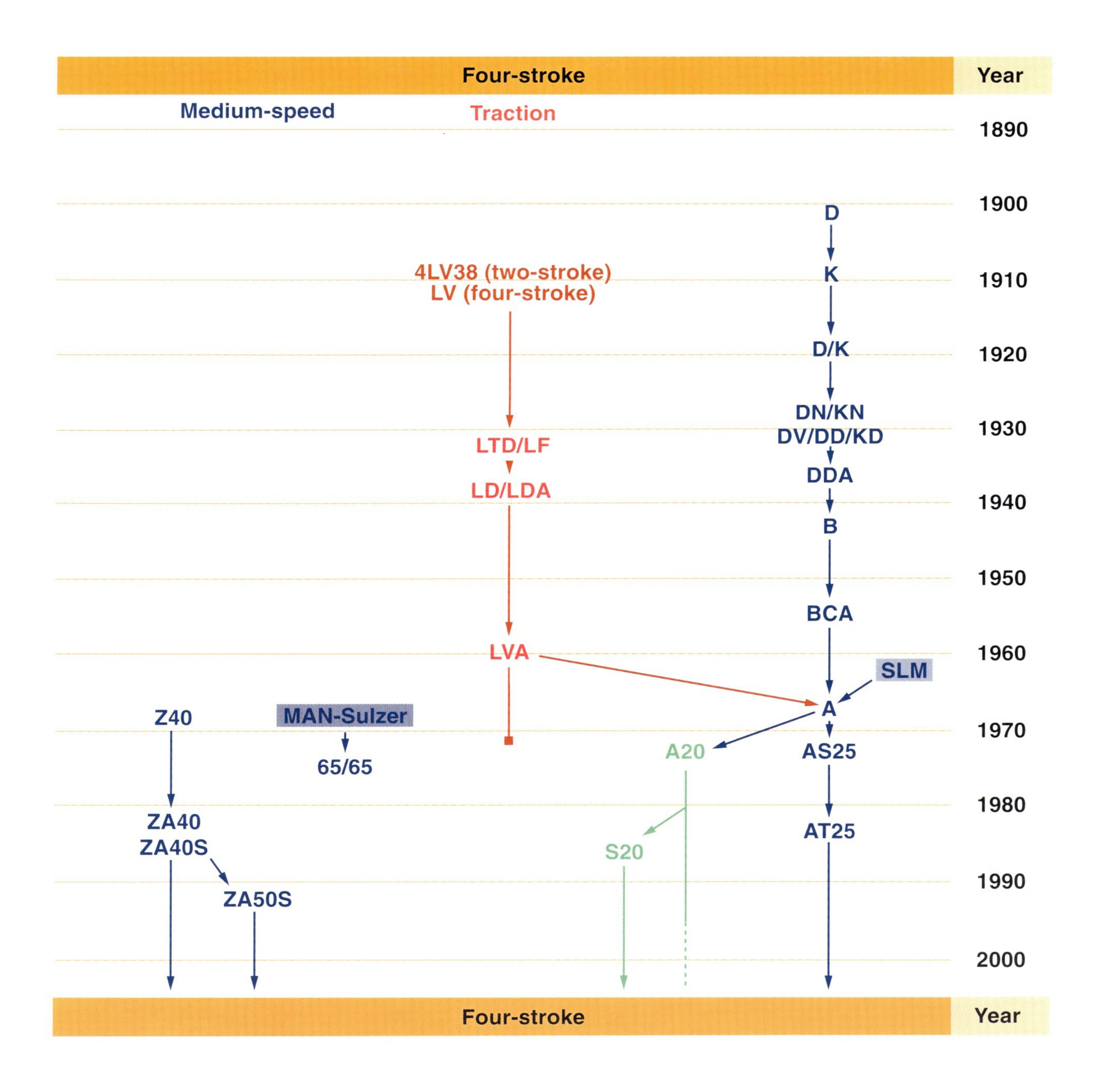

Principal parameters of Sulzer diesel engines

Principal parameters are given here for most of the specific engine models to which references are made in this book. It is not a comprehensive list of all Sulzer engine types but those included here can be regarded as representative for their times. It must be noted that some engine sizes were built with different piston strokes, and different cylinder outputs. Otherwise the power output quoted is for the maximum continuous rating (MCR), in other words the nominal rating. The mean effective pressure (MEP) and mean piston speed are calculated for that rating.

The initial number in the type designation denotes the number of cylinders in the engine and the next letters denote the engine series. Otherwise, particulars are given mainly without reference to the specific cylinder numbers. The powers for double acting engines (DS..., DZ... and DQ... types) are for the upper and lower cylinders combined. The full-load brake specific fuel consumption (BSFC) is only given when it is definitely known.

Year ordered	Type	Cycle	Bore mm	Stroke mm	Max Power hp/cyl	Speed rpm	MEP bar	Mean piston speed, m/s	BSFC g/bhph
1897	Series XVI	4	260	410	20	160	5.07	2.19	248
1903	1D40	4	310	460	40	200	5.08	3.07	210
1904	2D15	4	205	320	15	270	4.64	2.88	—
1904	2D20	4	230	350	22.5	260	5.25	3.03	240
1905	DM175/250	2	175	250	22.5	375	4.40	3.13	—
1906	3Z133	2	500	720	250	150	5.20	3.60	211
1907	4SNo.3	2	180	250	25	350	4.95	2.92	265
1909	4SNo.4	2	215	290	37.5	325	4.84	3.14	—
1909	4SNo.6a	2	310	460	95	250	4.83	3.83	—
1909	4LV38	2	380	550	400	304	9.31	5.57	—
1909	2LH30	2	305	380	125	350	5.68	4.43	—
1910	6U23	2	230	280	50	500	3.79	4.67	—
1910	1S100	2	1,000	1,100	2,000	150	6.81	5.50	—
1911	6U32	2	320	320	100	450	3.81	4.80	252.7
1911	4SNo.9a	2	470	680	212.5	160	4.97	3.63	210
1912	6Q37	2	370	380	150	380	4.26	4.81	—
1912	6Z300	2	760	1,020	625	132	4.52	4.49	208
1913	4S250	2	680	960	400	110	4.60	3.52	217
1914	6LV26	4	260	300	33.3	440	4.20	4.40	—
1914	4S36	2	280	420	72.5	260	4.76	3.64	256
1914	4S56	2	340	540	105	200	4.73	3.60	—
1914	1S54	2	540	500	520	390	5.14	6.50	183.5
1915	LF25	4	280	380	25	250	3.78	3.17	—
1915	6Q45	2	450	440	216.7	325	4.20	4.77	—
1917	4S60	2	600	940	312.5	100	5.19	3.13	188
1919	8Q54	2	540	570	375	300	4.23	5.70	195
1919	4ST60	2	600	1,060	337.5	100	4.97	3.53	—
1919	4ST68	2	680	1,100	500	100	5.06	4.00	—
1920	4D85	4	420	620	92.5	187	5.08	3.86	—
1922	6ST70	2	698.5	990.6	541.7	127	4.96	4.19	—
1923	4ST68	2	680	1,100	400	85	5.20	3.12	—
1923	6S76	2	760	1,340	600	90	4.84	4.02	183.4
1924	8Q65	2	650	680	812.5	280	5.67	6.35	—
1925	10ST68	2	680	1,200	580	110	5.34	4.40	177
1926	5S90	2	900	1,600	1,000	90	4.82	4.80	—
1926	DZ90	2	900	1,400	2,000	100	5.25	5.60	—
1927	6S47	2	470	820	400	156	5.30	4.26	—
1927	6Z68	2	680	960	550	136	5.12	4.35	—
1927	8ST68	2	680	1,000	1,000	120	5.06	4.00	—
1927	10ST76	2	760	1,340	700	100	5.08	4.47	167.5
1927	8ST82	2	820	1,440	880	100	5.11	4.80	177.5
1928	6ST78	2	787.4	1,092.2	666.7	115	4.81	4.19	—
1929	3Q51ES	2	510	550	500	390	5.04	7.15	159
1929	6SN48	2	480	900	250	150	4.52	4.50	160
1929	3DZ60	2	600	800	800	214	3.86	5.71	185
1929	8DSL70	2	700	1,200	950	106	4.54	4.24	—
1930	5DH38	4	380	600	100	250	5.19	5.00	—
1930	3DQD38	2	380	460	500	480	4.41	7.36	—
1930	9Q51N	2	510	550	444.4	390	4.48	7.15	—
1930	3DQ68	2	680	940	1,833.3	258	4.59	8.08	173.8

Year ordered	Type	Cycle	Bore mm	Stroke mm	Max Power hp/cyl	Speed rpm	MEP bar	Mean piston speed, m/s	BSFC g/bhph
1930	8DZL70	2	700	1,200	1,350	150	4.30	6.00	—
1931	8DZD60	2	600	1,000	925	187	3.86	6.23	—
1932	6F16	4	160	200	40	2,000	4.39	13.33	—
1932	6DDA22	4	220	320	61.7	500	8.95	5.33	162
1932	7SD56	2	560	840	478.6	225	4.54	6.30	—
1932	10SD56	2	560	840	430	215	4.27	6.02	—
1932	12SDT58	2	580	840	708.3	258	5.46	7.22	165
1932	6SD60	2	600	1,040	400	135	4.45	4.68	—
1932	8DZL76	2	760	1,200	1,425	136	4.25	5.44	—
1933	TS36	2	360	600	150	250	4.34	5.00	172
1933	6DSD53	2	530	760	600	215	3.15	5.45	—
1933	8SD72	2	720	1,250	687.5	126	4.77	5.25	149.2
1934	6LF19	4	190	230	48.3	1,200	5.45	9.20	—
1934	5SD36	2	360	680	150	240	3.98	5.44	173.5
1934	1QD42	2	420	500	400	450	5.66	7.50	—
1934	8SD49	2	490	640	375	288	4.76	6.14	—
1934	10DSDT76	2	760	1,200	1,400	135	4.56	5.40	168
1935	12LDA31	4	310	390	183.3	700	7.85	9.10	—
1936	6QD42	2	420	500	395	455	5.53	7.58	175
1936	3DQD42	2	420	580	750	450	4.58	8.70	—
1936	10SD58	2	580	840	650	250	5.17	7.00	—
1936	8SD65	2	650	1,200	450	110	4.53	4.40	158
1936	10STA68	2	680	1,200	735	120	6.20	4.80	—
1936	1SD72	2	720	1,250	723	120	5.22	5.00	—
1937	4ZGA19	2	190	2x300	354.5	750	12.30	7.50	143
1937	10QDCT51	2	510	600	690	420	5.91	8.40	178
1937	12SDT76	2	760	1,250	1,041.7	145	5.59	6.04	151.5
1937	6SDH45	2	450	600	300	280	4.95	5.60	—
1938	TS48	2	480	700	300	225	4.64	5.25	168
1938	10ZDA72	2	720	980	900	167	5.96	5.46	—
1939	6G18	2	180	2x225	260	850	11.79	6.38	—
1939	8SPG58	2	580	840	507.9	215	4.70	6.02	171
1939	1TS72	2	720	1,250	1,030	130	6.87	5.42	—
1939	11SD72	2	720	1,250	757.5	131	5.01	5.46	164
1941	M32	2	320	380	150	470	4.61	5.95	—
1941	6TA48	2	480	700	437.5	250	6.10	5.83	—
1941	7DSGS53	2	530	800	650	205	3.96	5.47	—
1941	6DSD60	2	600	1,000	700	135	4.05	4.50	—
1941	8SDS72	2	720	1,250	700	125	4.86	5.21	166
1942	6GA32	2	320	2x400	666.7	440	10.39	5.87	162
1944	8G18	2	180	2x225	312.5	1,000	12.04	7.50	175
1944	M51	2	510	550	400	320	4.91	5.87	—
1945	TD56	2	560	1,000	400	160	4.48	5.33	165
1946	7SDS60	2	600	1,040	450	135	5.00	4.68	—
1948	6SF72	2	720	900	750	185	4.88	5.55	—
1950	M42	2	420	500	275	360	4.87	6.00	—
1950	RSD58	2	580	760	520	240	4.76	6.08	—
1952	RSD76	2	760	1,550	1,000	119	5.27	6.15	—
1953	1RS58	2	580	760	500	230	4.78	5.83	—
1954	4TA24	2	240	400	105	400	6.40	5.33	—
1955	8VQA42	2	420	500	500	420	7.58	7.00	—
1955	RSAD76	2	760	1,550	1,300	119	6.86	6.15	155
1956	BCA29	4	290	360	150	500	11.14	6.00	—
1956	TAD48	2	480	700	440	250	6.13	5.83	—
1956	SAD60	2	600	1,040	683.3	160	6.41	5.55	—
1956	SAD72	2	720	1,250	900	125	6.24	5.21	—
1956	1RSA76	2	760	1,550	1,650	125	8.28	6.46	—
1956	RD76	2	760	1,550	1,700	122	8.75	6.30	—
1957	1RSA58	2	580	760	715	230	6.83	5.83	—
1957	RD90	2	900	1,550	2,300	122	8.44	6.30	—
1958	TAD56	2	560	1,000	500	155	5.78	5.17	—
1960	12LDA28-C	4	280	360	229.2	800	11.41	9.60	163.3
1963	ZV30	2	300	380	375	590	10.44	7.47	—
1963	Z40	2	400	480	550	430	9.36	6.88	—
1966	16LVA24	4	240	280	250	1,100	15.84	10.27	—
1966	A25	4	250	300	185	750	14.78	7.50	—
1967	RND90	2	900	1,550	2,900	122	10.64	6.30	153
1967	RND105	2	1,050	1,800	4,000	108	10.49	6.48	—
1969	6RD76	2	760	1,550	1,490	119	7.85	6.15	—
1970	Z40/48	4	400	480	600	500	17.56	8.00	—
1970	7RNMD90	2	900	1,550	2,900	122	10.64	6.30	—
1973	A20	4	200	240	125	1,000	14.63	8.00	—
1975	12V65/65	4	650	650	1,800	400	18.41	8.67	145
1976	AS25	4	250	300	270	1,000	16.18	10.00	158
1976	RND90M	2	900	1,550	3,350	122	12.29	6.30	149
1977	RLA90	2	900	1,900	3,350	98	12.48	6.21	145
1979	RLB90	2	900	1,900	3,600	102	12.89	6.46	140
1981	ZA40	4	400	480	870	600	21.22	9.60	139
1982	AT25	4	250	300	300	1,000	17.98	10.00	146
1982	RTA84	2	840	2,400	4,030	87	15.37	6.96	127
1985	ZA40S	4	400	560	980	510	24.10	9.52	136
1988	S20	4	200	300	197.5	1,000	18.50	10.00	143
1995	S20U	4	200	300	238	1,000	22.29	10.00	146
1995	ZA50S	4	500	660	1,630	450	24.67	9.90	133
1996	RTA96C	2	960	2,500	7,470	100	18.20	8.33	126

Main parameters of Sulzer two-stroke diesel engines

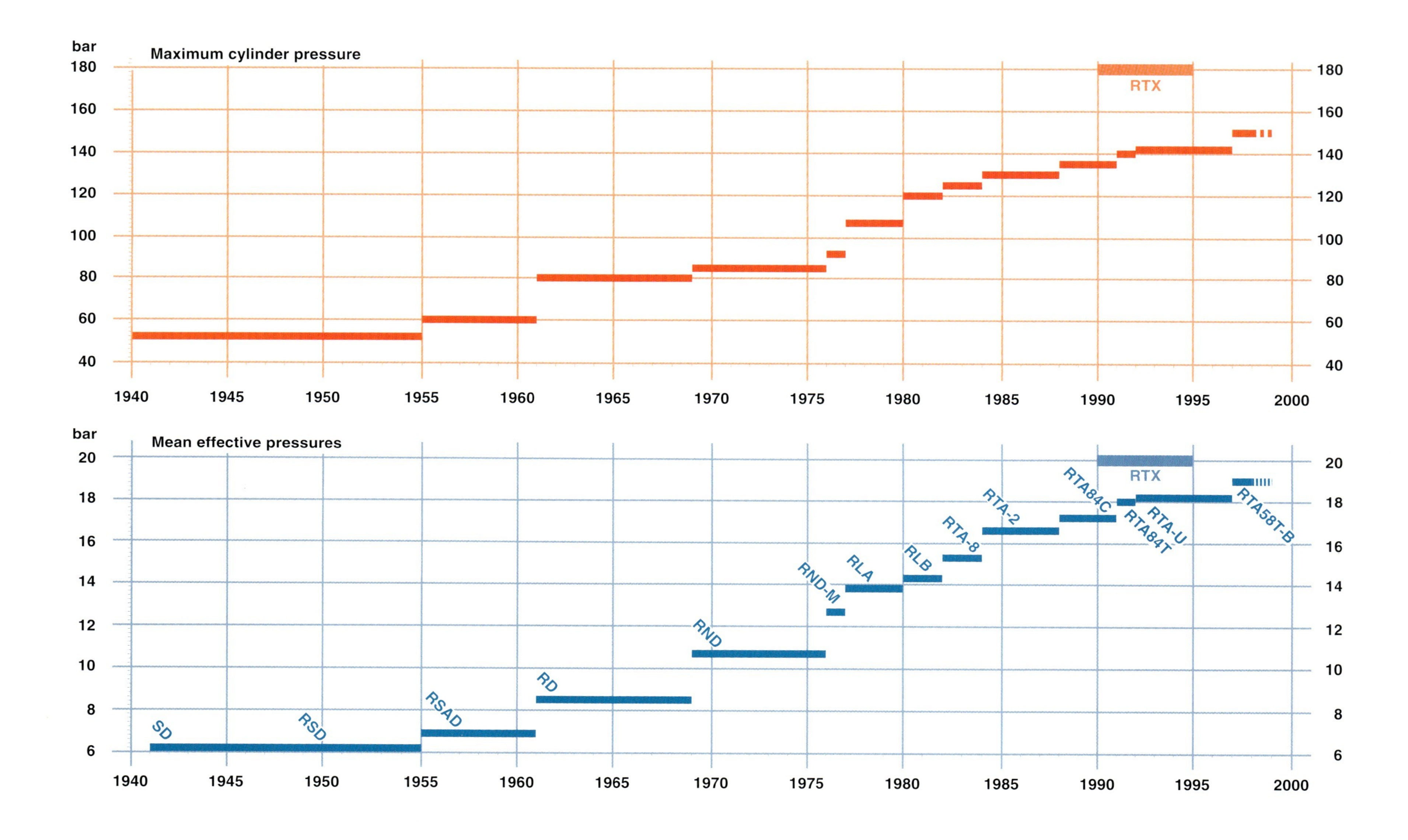

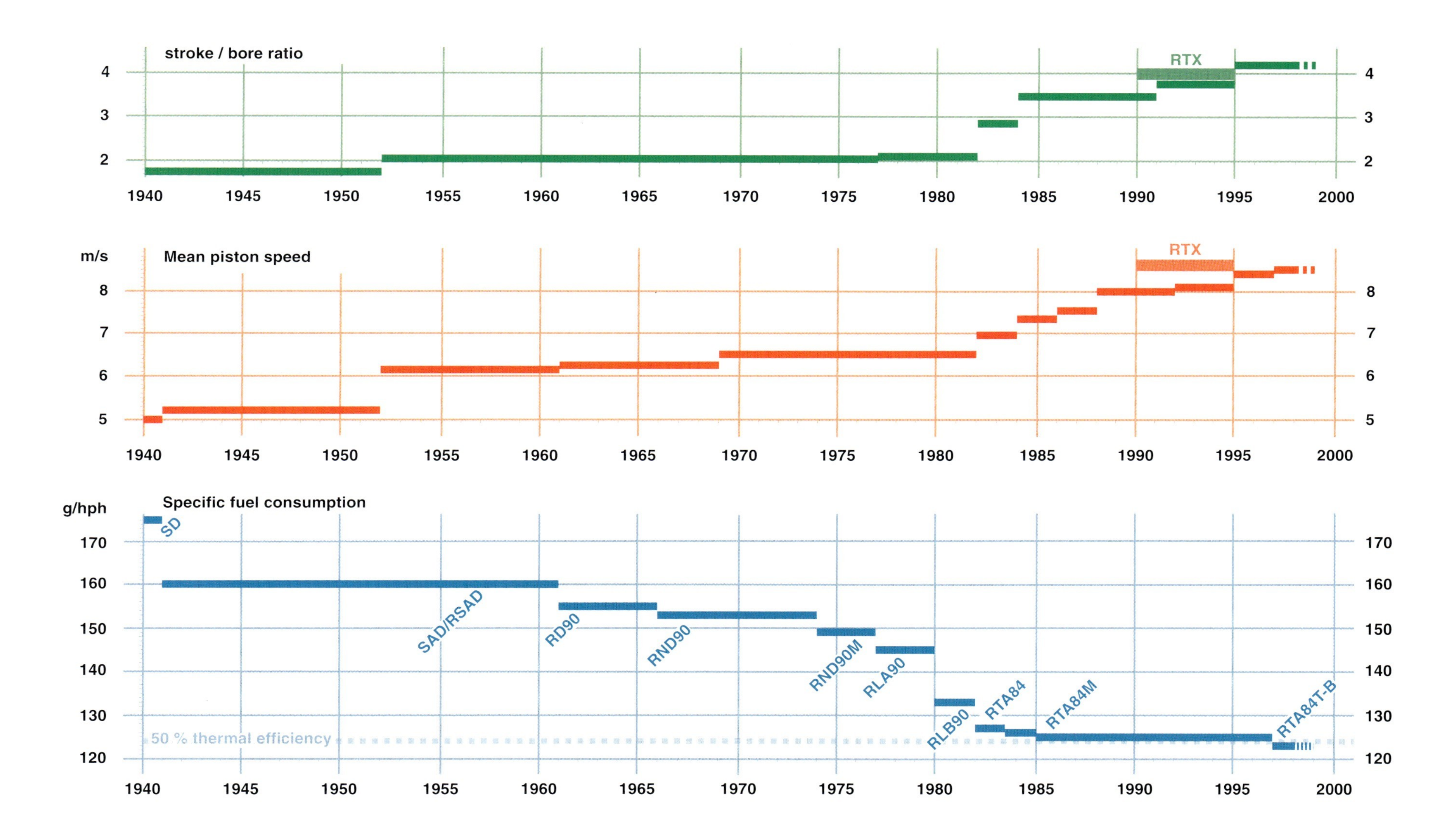
stroke / bore ratio
RTX
Mean piston speed
m/s
RTX
Specific fuel consumption
g/hph
SD
SAD/RSAD
RD90
RND90
RND90M
RLA90
RLB90
RTA84
RTA84M
RTA84T-B
50 % thermal efficiency

Past and present Sulzer diesel engine licensees

Licensee	Date
• Schweizerische Lokomotiv- und Maschinenfabrik (SLM), Winterthur, Switzerland (taken over by Sulzer Brothers Ltd in 1961).	July 1909
• L. Lang, Budapest, Hungary.	10 November 1909
• Société Anonyme des Forges & Chantiers de la Méditerranée (FCM), at Le Havre, La Seyne and Marseilles, France.	2 December 1910
• Busch-Sulzer Brothers-Diesel Engine Co, St. Louis, Missouri, USA (joint venture Adolphus Busch / Sulzer Brothers).	April 1911
• Erste Brünner Maschinenfabrikgesellschaft, Switzerland.	June 1911
• William Denny & Brothers, Dumbarton, Great Britain.	13 February 1913
• Howaldtswerke, Kiel, Germany.	28 June 1915
• Soc. Italiana Gio. Ansaldo & Co, Genoa, Italy.	16 August 1915
• Société Anonyme des Ateliers & Chantiers de la Loire (ACL), Paris, with works at Saint-Denis-sur-Seine, France.	26 October 1915
• A/S Thunes Mek. Verksted, Christiana (now Oslo), Norway.	25 January 1916
• Kolomna Machine Works (Kolomenskiy Maschinostroiteliy, Zavoda), Kolomna, Russia, for the Kolomna works and also the Sormovo works, Nishnij Novgorod.	16 February 1916
• Royal Norwegian Navy Department, Christiana (now Oslo), for Marinens Hovedverft (the Navy dockyard), Horten, Norway.	1 March 1916
• Soc. Española de Construcciones Metálicas, at Bilbao and Madrid, Spain.	29 March 1917
• Imperial Japanese Navy, Tokyo, Japan.	16 November 1917
• Cie de Construction Mécanique Procédés Sulzer (CCM), Paris, with works at Mantes, France. It was a subsidiary company of Sulzer Brothers founded on 27 December 1917.	January 1918
• Suzuki & Co Ltd, Kobe, Japan, for Toba Dock Co and Kobe Steel Works.	10 July 1918
• Sir W.G. Armstrong, Whitworth & Co Ltd, Newcastle upon Tyne, Great Britain.	18 September 1919
• Alexander Stephen & Sons Ltd, Linthouse, Glasgow, Great Britain.	22 March 1920
• The Wallsend Slipway & Engineering Co Ltd, Wallsend upon Tyne, Great Britain.	5 September 1920
• BV Koninklijke Maatschappij 'De Schelde' (Royal Schelde), Vlissingen, The Netherlands.	17 November 1921
• Northumberland Shipbuilding Co Ltd, Wallsend, Great Britain, with engines built at the subsidiaries: – Workman, Clark & Co Ltd (see below), – Fairfield Shipbuilding & Engineering Co Ltd (see below).	24 February 1922
• G & J Weir Ltd, Cathcart, Great Britain (for marine auxiliary engines only).	12 October 1922
• John Brown & Co Ltd, Clydebank, Glasgow, Great Britain.	10 January 1923
• G. Seebeck AG, Geestemünde, Germany.	27 February 1924
• Mitsubishi Zosen Kabushiki Kaisha (Mitsubishi Shipbuilding and Engineering Co Ltd), Tokyo and Nagasaki, Japan.	14 January 1925
• F. Schichau GmbH, Elbing, Germany.	18 June 1925
• Leningrad State Engineering Trust, Leningrad (now St Petersburg), USSR. Sulzer diesel engines were subsequently built at: – Komintern Locomotive Works of the Southern Engineering Trust, Kharkoff – Perviy Zavod Russkiy Diesel, – Nikolayev (André Marti) Shipyard – The Baltic Shipyard	1925
• Stabilimento Tecnico Triestino, Trieste, Italy, for the San Andrea Engine Works. The company was merged into Cantieri Riuniti dell'Adriatico (CRDA) when it was established in 1930.	7 September 1928
• NV Werkspoor, Amsterdam, The Netherlands.	12 December 1928
• Gebrüder Sulzer AG, Ludwigshafen am Rhein, Germany, as a subsidiary from 1881 to 1939.	February 1929
• Soc. Española de Construccion Naval, Madrid and Bilbao, Spain.	23 October 1929
• Workman, Clark (1928) Ltd, Belfast, Northern Ireland.	12 May 1931

• The Fairfield Shipbuilding & Engineering Co Ltd, Govan, Glasgow, Great Britain.	12 May 1931
• SA John Cockerill, Seraing, Belgium.	February 1932
• Odero-Terni-Orlando, Genoa, Italy, with factories in Sestri, Spezia and Livorno.	8 May 1934
• Sulzer Brothers (London) Ltd, London, Great Britain.	24 May 1934
• The Taikoo Dockyard & Engineering Co of Hong Kong Ltd, which was a subsidiary of John Swire & Sons Ltd, Great Britain.	27 September 1934
• American Locomotive Co (ALCO), Auburn, NY, USA.	12 July 1935
• R & W Hawthorn, Leslie & Co Ltd, Newcastle upon Tyne, Great Britain.	24 February 1936
• Cammell Laird & Co Ltd, Birkenhead, Great Britain.	16 September 1936
• Fritz Bührer, Bäretswil, Switzerland.	27 December 1939
• Dominion Engineering Works Ltd, Montreal, Canada.	16 July 1940
• Halberg, Maschinenbau & Giesserei AG, Ludwigshafen am Rhein, Germany.	October 1941
• Sociedad Española de Construcciones Babcock & Wilcox Compañia Anónima, Bilbao, Spain.	27 January 1942
• Gebr. Mägerle AG, Uster, Switzerland.	10 December 1942
• A. Johnson & Co, Stockholm, Sweden, for the subsidiary company A/B Karlstads Mek. Verksted, Karlstad, Sweden.	2 April 1946
• Richardsons, Westgarth & Co Ltd, Middlesborough, Great Britain, with engines built by two subsidiaries: – North Eastern Marine Engineering Co (1938) Ltd (NEM), – George Clark (1938) Ltd, Sunderland, which merged together in 1964 to form George Clark & NEM Ltd, Wallsend on Tyne, which subsequently merged with Hawthorn Leslie (Engineers) Ltd (see above) to become Clark-Hawthorn Ltd. It was later combined with Kincaid Ltd, Greenock, to form Clark Kincaid Ltd and finally changed to Kvaerner Kincaid Ltd.	3 December 1946
• La Maquinista Terrestre y Maritima SA, Barcelona, Spain.	30 May 1947
• Empresa Nacional 'Bazan' de Construcciones Navales Militares SA, Madrid, with works at Cartagena, Spain.	10 March 1947
• The Harima Shipbuilding & Engineering Co Ltd, Aioi, Japan, which was merged with Ishikawajima in 1960 to form Ishikawajima-Harima Industries Co Ltd (IHI), Tokyo, with the works at Aioi. The enginebuilding activities of IHI and Sumitomo (see below) were transferred to the joint company Diesel United Ltd established on 1 October 1988, with works at Aioi.	13 November 1948
• Uraga Dock Co. Ltd, Tokyo and Tamashima, Japan, becoming Uraga Heavy Industries Ltd in 1964 which was merged in 1969 with Sumitomo Machinery Ltd to form Sumitomo Shipbuilding & Machinery Co Ltd and later Sumitomo Heavy Industries Ltd, Tokyo. In 1988, Sumitomo enginebuilding activities were transferred to Diesel United Ltd (see above).	8 March 1950
• Henschel & Sohn GmbH, Kassel, Germany.	24 October 1950
• Barclay, Curle & Co Ltd, Glasgow, Great Britain.	November 1951
• Swan, Hunter & Wigham Richardson Ltd, Wallsend upon Tyne, Great Britain.	November 1951
• Brodogradiliste i Tvornica Dizel Motora '3 Maj', Yugoslavia, which, in 1991, became '3 Maj' Engines & Cranes, Rijeka, Croatia.	10 January 1954
• Wärtsilä-Koncernen A/B, Vasa, Finland, with works at Turku, Finland.	1 October 1954
• Jugoturbina, Tvornica Parnih Turbina i Dizel Motora, Karlovac, Yugoslavia, which, in 1992, became Adriadiesel DD, Karlovac, Croatia.	28 March 1955
• Société Anonyme des Chantiers de l'Atlantique, Paris, France, with works at Chantiers et Ateliers de Saint-Nazaire Penhoet.	1 October 1955
• 'Masinimport', State Enterprise for External Trade, Bucharest, Roumania for the works of Combinatul Metalurgic, Resita.	26 April 1956

• Centromor (Centrala Morska Importowo – Eksportowa, PP), Warsaw, Poland, (later the Foreign Trade Office H. Cegielski, Poznan) for the works of: – Zyklady Przemyslu Metalowego H. Cegielski, Poznan (HCP), – Zyklady Urzadzen Technicznych 'Zgoda', Swietochlowice, – Stocznia Gdanska im Lenina, Gdansk, – Puckie Zaklady Mechanizne, Poland.	6 September 1956
• Iino Shipbuilding & Engineering Co Ltd, Tokyo, Japan, which later became Maizuru Jukogyo Ltd and in 1966 was taken over by Hitachi Zosen Corp, Osaka, with works at Osaka, Maizuru & Sakurajima.	26 November 1956
• Vickers-Armstrongs (Engineers) Ltd, Barrow-in-Furness, Great Britain with the subsidiaries: – Canadian Vickers Ltd, Montreal, Canada, – Cockatoo Docks & Engineering Co Pty Ltd, Sydney NSW, Australia.	1 May 1957
• Maschinenfabrik Buckau R. Wolf AG, Kiel, Germany.	1 October 1957
• David Rowan & Co Ltd, Glasgow, Great Britain.	15 January 1958
• Empresa Nacional 'Elcano', which merged in 1965 with Astilleros de Cadiz SA, Madrid and Valencia, Spain. The company later came within the state shipbuilding group Astilleros Espanoles SA (AESA). On 1 November 1994, the agreement was transferred from AESA to Manises Diesel Engine Co SA, Valencia, Spain.	17 November 1959
• Scotts' Shipbuilding & Engineering Co Ltd, Greenock, Great Britain.	18 March 1960
• Astilleros y Fabricas Navales del Estado SA (AFNE), Buenos Aires, Argentina, with works at Rio Santiago.	July 1960
• Commonwealth of Australia for the Commonwealth Government Engine Works, Port Melbourne, Victoria, Australia.	9 March 1961
• Marinens Hovedverft (the Navy dockyard), Horten, Norway.	20 April 1961
• Ishikawajima do Brasil-Estaleiros SA (Ishibras), Rio de Janeiro, Brazil.	8 June 1962
• Nordberg Manufacturing Co, Milwaukee, Wisconsin, USA.	1 September 1964
• English Electric Co Ltd, London, Great Britain.	1 May 1967
• Koraboimpex, Varna, with works at KTM Russe, Bulgaria.	30 July 1973
• Taiwan Machinery Manufacturing Corp (TMMC), Kaohsiung, Taiwan, Republic of China.	8 July 1974
• Hyundai Engine Manufacturing Co Ltd, Ulsan, Republic of Korea, which later reverted to the parent company Hyundai Heavy Industries Co Ltd, as the Engine and Machinery Division.	11 June 1975
• Westinghouse Electric Corporation, Pittsburgh, PA, USA.	15 June 1976
• Ssangyong Heavy Industries Co Ltd, Seoul, with works at Changwon, Republic of Korea.	20 December 1977
• China National Technical Import Corp (later China State Shipbuilding Corp), Beijing, China PRC, for the works of: – Dalian Shipyard, Dalian – Dongfeng Shipyard, Chongqing – Shanghai Shipyard, Shanghai – Sichuan Diesel Engine Works, Chongqing – Yichang Marine Diesel Engine Plant, Yichang – Hudong Shipyard, Shanghai, added to agreement in 1995.	17 July 1978
• Sulzer Brothers, Inc, New York. (Licence to manufacture engines in the USA through subcontractors.)	29 June 1979
• Waukesha Engine Division of Dresser Industries Inc, Waukesha, California, USA.	1980
• Türkiye Gemi Sanayii AS, (Turkish Shipbuilding Industry Inc.) Istanbul, with works at Pendik, Turkey.	1981

• Diesel Ricerche SpA, Trieste, for the works of Fincantieri Cantieri Navali Italiani SpA, Diesel Engines Division, Trieste, Italy, for Sulzer RTA and Z engine types. On 1 July 1994, the agreement was transferred from Diesel Ricerche to Fincantieri Cantieri Navali Italiani SpA, Diesel Engines Division, Trieste. On 1 January 1997, the Trieste works was formed into a separate company Grandi Motori Trieste SpA as a subsidiary of Fincantieri.	1982
• Ito Engineering Co Ltd, Shimizu, Japan. (Sub-licence from Sumitomo Heavy Industries Ltd for A-type engines).	1 April 1983
• Nippon Kokan KK (NKK Corporation), Tokyo, with works at Tsurumi, Japan.	1983
• VEB Dieselmotorenwerk Rostock (DMR) which, in 1990, became Dieselmotorenwerk Rostock GmbH, Rostock, Germany.	1983
• Korea Heavy Industries & Construction Co Ltd (Hanjung), Seoul, with works at Changwon, Republic of Korea.	1984
• Bremer Vulkan Verbund AG, Bremen, Germany.	1991
• Fincantieri-Cantieri Navali Italiani SpA, Diesel Engines Division, Trieste, Italy, for the works of Diesel Engines Division, Bari, Italy, for Sulzer S20-type engines. In 1996, the Bari works was formed into a separate company Isotta Fraschini SpA as a subsidiary of Fincantieri.	7 January 1994
• Samsung Shipbuilding & Heavy Industries Co Ltd, Seoul, with works at Changwon, Republic of Korea.	21 March 1994

The Sulzer D

SULZER Diesel
SULZER Diesel
SULZER Diesel
SULZER Diesel
SULZER Diesel
SULZER Diesel
SULZER Diesel
SULZER Diesel
SULZER Diesel
SULZER Diesel
SULZER Diesel
SULZER Diesel
SULZER Diesel
SULZER Diesel
SULZER Diesel
SULZER Diesel
SULZER Diesel
SULZER Diesel
SULZER Diesel
SULZER Diesel
SULZER Diesel
SULZER Diesel
SULZER Diesel
SULZER Diesel